Where's Meriden?
The Demise of Small Town U.S.A.

TWIN CITIES

MERIDEN

By the same author

THE DAY OF THE BONANZA

THE CHALLENGE OF THE PRAIRIE

BEYOND THE FURROW

TOMORROW'S HARVEST

KOOCHICHING

PLOW SHARES TO PRINTOUTS

TAMING THE WILDERNESS

HISTORY OF U.S. AGRICULTURE
(a.k.a. Legacy of the Land)

CREATING ABUNDANCE

E. M. YOUNG: PRAIRIE PIONEER

A CENTURY IN THE PARK

PRAIRIE HOMESTEAD TO WALL STREET

WHERE'S MERIDEN?
The Demise of Small Town U.S.A.

A microstory of the rise and decline of a small town impacted by
free-enterprising farmers who helped industrialize agriculture as it became
a key to making America the world's most powerful nation

HIRAM M. DRACHE
Concordia College
Moorhead, Minnesota

HOBAR

HOBAR PUBLICATIONS
A Division of Finney Company
www.finney-hobar.com

ISBN: 978-0-913163-47-4

Designed by Terin Martin
Edited by Keri Stifter
Indexed by Dianna Haught, Words by Haught

Hobar Publications
A Division of Finney Company
8075 215th Street West
Lakeville, Minnesota 55044
www.finney-hobar.com

1 3 5 7 9 10 8 6 4 2
Printed in the United States of America

About the Author

Hiram Drache is a native of Meriden, Minnesota. After serving in the Air Force as a navigator on a B-17 bomber in World War II, he earned degrees from Gustavus Adolphus College, the University of Minnesota, and the University of North Dakota. During his post-graduate years he farmed, worked in several businesses, and taught high school. He purchased his first farm in 1950 and was involved in farming until 1981, when he sold his family farm and leased out the others. He has taught at Concordia College in Moorhead, Minnesota, for all but two years since 1952, and since 1991, he has been Historian in Residence. During those early years, prior to 1991, he wrote his first six books and made over a thousand speeches in thirty-six states, six provinces of Canada, Australia, England, Germany, and Norway. This is his thirteenth book. In addition he has contributed to eight others and has written more than fifty articles chiefly on contemporary agriculture and agricultural history.

About the Cover Artist

Sandy Dinse was born and raised in Steele County, Minnesota. While attending a one-room schoolhouse, she remembers looking forward to art on Fridays. The rural upbringing and living on a family farm has been a source of inspiration for a way of life she cherishes along with her love of history that is evident in her paintings. Through several of her watercolors she strives to bring about an awareness of how important the family farm is to so many people.

After her marriage in the early 1960s she and her husband farmed in Meriden Township. At that time the village had a full complement of Businesses, most of which were operated by the families who owned them. Many of the residences were occupied by retired farmers whose families were operating the farms.

As her family grew she found more time to pursue her artistic interests. This included enrollment in workshops taught by gifted artists which brought many benefits to her art. Her nostalgic nature was her motivation to design and construct metal and wood sculptures from the charred ruins of three area churches destroyed by fire. One was her own parish.

Later life has brought retirement, continued love of the land, church, community and the never ending passion for painting.

About the Cover

The essence of the painting is to bring Meriden back to life as it was in its prime. It is the hope of the artist that the viewer, if listening carefully, can hear the clang of the blacksmith's hammer, the ringing of the church bell, the rumble of the wagons and the sound of the horses' feet.

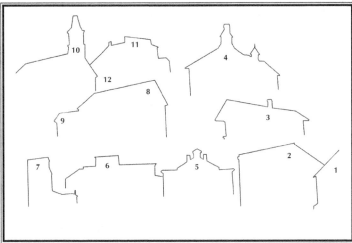

1. The second schoolhouse which was moved adjacent to the bank and became the barber shop/pool hall.

2. The bank shortly after it was built in 1915.

3. The CNWR depot.

4. The creamery in 1890.

5. The schoolhouse 1922-2010.

6. Hayes Lucas hardware and lumber shed on left.

7. The elevator in 1924.

8. The blacksmith shop with Grandprey Hall on second floor at its original site prior to 1915.

9. The south end of the living quarters of the pool hall.

10. The original St. Paul's Lutheran church 1976-1936.

11. The Central Hotel upper floor and store 1879, after 1964 the Kath Store and attached café, later office for the propane business.

Foreword

I can think of few guests I interviewed over my thirty plus years at KRFO Radio in Owatonna who inspired me as much as Dr. Hiram Drache, the author of this book. There are thousands of histories that have been written about all sections of the United States, but few that I have read have offered the personal perspective from the author's viewpoint as the one you are about to read. When Dr. Drache was my guest on a number of radio programs, I found that he was not hesitant to speak his mind on a number of subjects . . . ranging from farming to development of the various rural communities of the nation including Steele County and Meriden Township. The fact that he grew up in this township enables him to share stories of the many personalities that shaped its history. His book offers a fascinating description of Meriden as he takes you on a chronological journey from Meriden's early days to the present with emphasis on agriculture. It is a sound story that can only be told by a person who actually lived there and shared the good times and the bad, which he in turn will share with you. Enjoy this fascinating piece of rural Americana and Steele County only as Dr. Drache can tell it. It is a trip back in time that you will totally enjoy and that may ultimately help you understand our present and how we arrived here.

Todd Hale
Former Vice President
Owatonna Broadcasting Company

Preface

History has always been my favorite subject, and I enjoy doing research. In 1961 when I first thought about a history of Meriden Township, I knew that I would not use the typical format where a commercial complier got people to write about their families–when they came, settled, passed on, and how many years they had served on the school and township boards, and the church council. There was always a family picture, but there was never anything included about the world beyond the township borders. Except for a brief period in 1947, I have not lived in the township since 1942, but whenever I returned for a visit, I interviewed some of the older residents.

By 1983 I had written five books, and my mother hinted that I should write a local history and gave me notes about early events. After she died in 1988 I gave a local history more thought and wrote a fifty-two page family history that had items about the community in it. As part of his record keeping for the trucking business, Dad had kept a day book for each year from 1937 to 1988 in which he often inserted data of historical interest about the community.

By 1992 I had accumulated a file drawer full of letters, newspaper clippings, obituaries, pictures, and miscellaneous notes. The notes showed several major themes: family, village, township, county, and how changing agriculture caused the loss of the small businesses in the village. It was then that I realized that I should write a micro-macro history of what happened to my home town and why it happened.

On June 10, 1992, I wrote to thirty-three people I had known when I lived in the township and asked them to write about their memories. In the letter I stated that unless they could guarantee that they would live another ten years that they should write their memories now. You can imagine how thrilled I was when I received responses from four, all of whom no longer lived there, but they all had some interesting stories and had good feelings about their early life there. By then I had attended enough high school, college, and organization reunions to realize that often those who lived across the street from the reunion site never attended, and probably they felt the same way about a written history.

My parents settled in Meriden after their marriage and always had both the Owatonna and Waseca telephones. Partly because of that, and the fact that my mother had commercial training and three years of business experience, she became a correspondent for one of the weekly papers in each of those towns. She gathered and sent the same news to both. The papers sought to have a correspondent in every neighborhood of the township. Difficult travel conditions restricted social contacts to "neighborhood" groups, which formed around ethnic, religious, economic, or educational preferences. Over the years seven designated "neighborhoods" had developed in the township, but as transportation improved, the "neighborhoods" lost their identity and the tenor of the news changed, but the weekly papers were still the best source of obtaining the township news.

James A. Child's *History of Steele and Waseca Counties*, published in 1887, proved to be a good starting point for this story. For the lack of any other histories, I reasoned that the weekly newspaper was the best source of information, so I chose the *Owatonna Journal* (also called the *Weekly*, *Herald*, and *Chronicle*), which was published from January 1868 to January 1938. It was succeeded by the *Steele County Photo News*, then the *Photo News*, which was published from 1938 to 1975 and focused on town and rural events. But as agriculture changed, the rural areas changed, and people lost their interest in the "neighborhood" news as their horizons broadened; as a result the paper could no longer sell advertising to remain solvent.

The history of the village is a typical story of thousands of little agricultural service centers that once dotted the nation, but many no longer are here. One and one half miles south of the village is one of the enduring symbols of the township: St. Paul's Lutheran Church. I did not include much about the church because there are several histories available. I have included little information about the schools because there is a county history of what the rural schools were like. Nor is there much written about the three township creameries because Bernie Hanson wrote *History of the Creameries of Steele County*, which covers that topic well. There is only scattered information about R. James Stuart, the renowned artist, because his story is covered in *The Life and Works of R. James Stuart* by Sterling Mason.

If this story has more about the Drache family than you care for it is because during research I learned how deep my roots really are in the township. My great, great grandfather was Christopher H. Wilker, who came to the township in 1856; my great grandmother was the first white child born in the township; and my parents were involved in village business affairs for about fifty years.

When I started research I had no idea at what point the story would close, but changing agriculture, which I have written about in several books, dictated a new era for rural society. Instead of having a farm service center every eight miles to accommodate horse and wagon, by the 1960s thirty-five minutes (not miles) was suggested to be the ideal distance to justify an adequate trade center. The farmers of Meriden, Steele County's most pro-ductive township, had led the way to the demise of what Thorstein Veblen once called "the country town a major and enduring part of our culture . . . one of the most wasteful institutions ever to exist."

In my research I read a book written in 1929 entitled *Too Many Farmers*, by Wheeler McMillen, a sage of American agriculture who I had the privilege to interview more than fifty years after he wrote that book. The book pointed out that there were "more people in agriculture than are able to derive an American standard of living. . . . More people are farm-ing than can make decent profits with the prevailing levels of prices and costs." Then he wrote that improving technology was steadily increasing the productivity of the farm worker, which meant that fewer farmers would be needed in the future. "This should be. The fewer people [who] have to be engaged in producing the necessities of life, the more are freed to aid in producing the luxuries, the conveniences, the materials and services that create and enrich civilized existence." The tempo of industrialization in agriculture continued. Tractors and accompanying equipment grew ever larger, which enabled farmers to till more acres. The countryside was over-capitalized with an excess labor force and too many farms.

I had studied the USDA's research on large farms of the 1920s so I knew that its economists knew what was taking place, but I was amazed when I learned that some of the leaders of the Gavel Club of Owatonna, which was founded in 1928, had the same vision about changing agricul-ture. They realized that the farming community would be a source of la-bor in the years ahead. In recent years more than 6,000 people from the surrounding countryside commuted to Owatonna to work and more than 60 percent of the income of the county was derived from their labor. The former "Butter Capital of the World" no longer existed, for Steele County had become a non-farming dependent county. By 1988 less than 10 percent of its total earnings came from agriculture, but it was the only business of Meriden Township. Nationwide, 20 percent of the farmers were producing 85 percent of the total output.

The township was settled before the Homestead Act was passed, so people who came here had to pay for the land or at least have sufficient credit to secure financing. Unfortunately, many of the homesteaders had little knowledge of farming or business. This partially explains why those who broke the sod here did not experience the failure rate as the homestead-ers who secured land from the government by fulfilling the homestead re-quirements. I have often stated that no act of a well intended and benevolent government has been the cause of so much hardship as the Homestead Act, but it did hasten the settlement of our nation.

Many of the township's original land owners were speculators who hoped for a quick profit through resale. Of the 121 names recorded in the Book of Original Entry up to 1858, only 18 remained as title holders on the 1879 plat book. A few of the first settlers later retired and settled in Meriden, but others moved to Owatonna or Waseca.

In 1955 we sold a farm in Somerset Township that we had owned since 1950 because we thought we would be permanently living in Steele County, and purchased another near Concordia College in Moorhead. Harry Winters, our accountant, purchased the Somerset farm. I asked him why he wanted to own a farm. He replied that in his business he had observed that "farmers lived poor but died wealthy." When I visited with him nearly fifty years later he said buying that farm was the best investment he ever made. During the years covered by this history the *Consumer Price Index* indicates that the basic prices have increased 23 fold, but land prices have multiplied 3,200 fold. This explains his observation. Fortunately, over the years farmers have profited by the long-term rise in land values. However, they have become more business oriented and realize that they had to have a large enough farm so they did not have to deny themselves of a decent living and still maintain a viable business operation.

The really enduring symbol of the township that survives is a country side of fertile fields in a very different form from what it was in 1855 but far more productive. The Morrill Act of 1862 provided for a system of agricultural education, research, and extension that is the envy of the world and has given us an industrialized agriculture that provided the backbone for the world's leading industrial nation.

Meriden Township was consistently a top producer in the county because of the fertility and lay of the land. The Crane Creek drainage system improved on what nature provided. A core of progressive farmers in the township, and in all townships, inadvertently caused the decline of small town U.S.A. with the help of government programs.

Farming became a commercial enterprise and farmers had to go to Owatonna or Waseca or whatever their county seat towns were to do their business because it took larger firms to offer the services they required. The small towns, like farming, were overcapitalized with both small businesses and labor. They filled a niche when they were needed, and today they live on in nostalgia, in poetry, in stories about the "good old days," and in paintings about the rustic past, which I thoroughly enjoyed because I knew no better then but I am not sure any of my grandchildren would enjoy today. Bob Bergland, Secretary of Agriculture, 1978-1982, told me that the USDA had spent millions in an attempt to save the small towns and had nothing but grass to show for it.

Those readers who are interested in more details than I have included in the text can turn to the Appendix. Nearly all of that data came from the local paper supplemented by data from the USDA. The prices listed are what the farmer received. The economic data justifies Thorstein Veblens's view of the small towns of his time, but how else would those who broke the sod have survived?

There are many tables in the Appendix that point out how the change in the value of the dollar over the years has benefitted those who had the risk orientation necessary to be a farmer. For a better understanding of some of the changes that have taken place since 1855, I have inserted this *Consumer Price Index* table based on 100 in 1860.

1860—100	1920—2401	1960—354	1990—1,563
1890—109	1930—2001	1970—464	2000—2,059
1900—101	1940—1681	1980—985	2010—2,275
1910—114	1950—288		

Use the *Consumer Price Index* to your advantage. Other data gives a list of the teachers whose names were published in the paper. There was an inconsistency in the spelling of some of the names, which probably came about because the correspondent for the paper did not get the correct spelling or their handwriting was not legible.

As agricultural economist, Earl Heady, stated in 1980, "If we are to have some small farms left, because policy-wise we decided that we should have small farms, then we should know how to tackle the small farm problem and guarantee that they will have income enough to bring their families opportunities that exist for other families in our society." It is my hope that the reader will enjoy this story of my home town and understand the part that it and other towns like it played in the development of rural America from its primitive agriculture to the highly sophisticated industry that it is today.

-Hiram M. Drache

Acknowledgements

To Ada, the girl from the farm whose sparkling eyes attracted my attention the first month that she was in college. Her college career only lasted three years, but ours has lasted over sixty-four years. Little did I know then that she understood farming, liked history, was an excellent grammarian, and a good speller. Commencing with my first Master's paper in 1950 ,she has always been my primary editor. This is our sixteenth book-length manuscript in addition to many chapters and articles for other publications.

I owe much to the Journal Chronicle Company, an Owatonna institution that published newspapers from 1868 to January 1938, which provided me with 1,497 research notes that in some respects are the backbone of this volume. When the *Journal* was discontinued, the *Steele County Photo News* tried to fill that void and provided me with 430 notes to cover from June 1938 until 1975 when changing times forced it out of business.

Sandy Dinse and the late Mrs. Frank Fererichs collaborated to translate the minutes of St. Paul's Lutheran Church from German to English. This was a great help because the correct date of birth, confirmation, marriage, and the spouses' name and data could be verified. I have probably used those records more than anyone since their translation. Read the blurb about the book cover by Sandy. I doubt whether many townships will have such an attractive and descriptive cover.

Kenneth Dinse answered many questions about current township activities and in the process we learned that C. H. Wilker, who is my great-great grandfather, is his great-great-great grandfather. Kenney's uncle, Norbert Abbe, and I shared the same birth year, and he was a favorite boyhood pal.

Kenny and James M. Andrews, both natives of the township and current town board members, were able to answer all my questions about events that were not otherwise available. The Andrews farm is the longest owned farm in the township and the Dinse family is the second longest. I knew their families well.

Laura Resler, director of the Steele County Historical Society, for her helpful hints, and Nancy Vaillancourt, of the Society, who located and researched the minutes of the Meriden Rural Telephone Company.

Jennifer McElroy of the Minnesota Historical Society—thanks for finding historic maps, as well as unique photos, which are not otherwise available. The Society has been collecting since territorial days.

Laura Ihrke and Lori Johnson for providing leads as to where to locate the unique tidbits that help add flavor to the story, and Nick Flatgard for the great photo. Allen Radel, whose ancestors I remember from my days in 4H, for his input relative to recent land transactions.

A special thanks to the late Arthur Uecker, a great friend of the Drache family dating to 1935, who provided me with a tape recording of his years as the Meriden butter maker as well as the minutes of those years. He also provided me with his insights as a leader in the final years of the dairy processing industry in Steele County.

Thanks to all who consented to be interviewed dating back as early as 1972. Ironically, most of the thirty-three who I wrote to in my first letter did not record their memories and did not live to provide an interview. History is so fleeting.

It was fun going to work at the Concordia College library every day to be met with these smiling people, in the order that I saw them: Bonnie, Suzanne, Linda, Jennifer, Leah, Carol, and Mary. In addition to their smiles there was often something available for my sweet tooth. If there was a mechanical problem, I could call Bonnie; if I had a problem that was really serious, I would drop in on Sharon; and when I got to fourth floor Liza was always there with a smile and something from the archives that she thought I might want to use. Life is fun, but when there is a problem with the computer I called 3375, computer support, and Bruce, Mitch, Jon, or one of their associates always were ready to help. They eagerly greeted that "author" who has used computers longer that most of them are old but still had to call on the "computer experts" for help.

I can never forget that my parents instilled a love of history and geography in me and lived to see and read the first six of my books. Their satisfaction was all the payback I needed for the many spankings they had applied in my earlier days.

Table of Contents

CHAPTER I
The Setting 1855–1869

Early Settlers

At the time of this writing Meriden is only 155 years old. The United States was founded 225 years earlier and the first European immigrants had arrived 403 years previously. This gives a proper perspective on where Meriden fits in the realm of history.

Until the 1950s the law of primogeniture in Scandinavian countries provided that the eldest son would inherit the parents' property, which meant that the remaining children faced a life of working in the forest, becoming a fisherman, joining the military, or leading the life of a day laborer. The industrial revolution was just starting to make its impact on the other Western nations, but the options were not much better. By the mid-nineteenth century those nations were going through a period of political turmoil and wars, which caused many of the lower classes to realize that they had limited opportunities to improve their future. They heard about free land, the great magnet, in America, which caused "America fever," and millions left in search of the American dream. Germans, Norwegians, and Swedes came in search of land while Irish and Italians sought urban opportunities. During the century after 1825, more Norwegians came to America than were in Norway in 1825. The choice farm land of mid America appealed to them. Steele County, and in particular for this story, Meriden Township, attracted them. Except for three of the northern sections and a scattering in two southeastern sections of the township, there was little woodland to retard breaking the sod to begin farming. Bradley Lake and Crane Creek were the only sizable water bodies to hamper them.

The Land Ordinance of 1785 made the great public domain available at first for $1 an acre and, in 1796, $2, causing land speculation to border on being a national craze. In the 1820s the purpose of selling the land was changed from raising money for the government to meeting the needs of the settlers. The price was established at $1.25 an acre for plots as small as eighty acres and on easy terms, which made it available to most settlers. Homesteaders and small-scale speculators were pleased when they did not have to bid against better financed investors who could afford to pay a higher price, and the land craze continued. The initial population of the township appeared quite fluid, for according to the Book of Original Entry, about 85 percent of those who initially acquired land in the township were no longer there by 1879, based on the owners listed in the plat book of that year.

In the summer of 1853, A. L. Wright and L. M. Howard and three others came and staked out claims in what became Medford Township. Howard and Wright erected a cabin and spent the winter of 1853-54 in the area. In 1854 Congress funded stage lines to provide mail service by giving them one section of land for each twenty miles of stage route a company operated. In 1855, to encourage immigrants, the stage was allowed to provide passenger service. The first passenger stage arrived in Owatonna in June, and for a brief period Owatonna was the terminus. However, shortly after, the stage company decided to extend its route from Winona to St. Peter, and in September the terminus was moved northwest of Owatonna to section six on the bank of Crane Creek. That east-west stage operated until 1866 when the railroad arrived.

When Steele County was first organized on February 20, 1855, it included all of what became Waseca County and what were the two western tiers of townships in Steele County. But in 1856 the state reorganized the counties, removed the western townships that became Waseca County, and took Merton, Havana, Auroa, and Blooming Prairie townships away from Dodge County and put them in Steele, making the county as it is today. [1]

In June 1855, A. M. Fitzsimmons staked his first claim in section thirty-six in the extreme southeast quarter and is credited as the first settler in Meriden Township. It appears that the stagecoach traveled to Wilton, which at that time had visions of being the county seat for Waseca County. Fitzsimmons had a stage stop on his land. This possibly could have been the location of what was called East Meriden, and may have handled mail. I have not been able to determine the exact location of that post office. None of the settlers' names are on property records, but the Post Office Department verifies that Marcus C. Flower was postmaster from December 22, 1856, until Joseph Guenther took over December 18, 1860, who was followed by Frederick J. Stevens on February 12, 1861. Stevens was active in politics at the township and state levels and gave the township the name of Meriden after his former home in Meriden, Connecticut. He was followed on January 12, 1869, by Archibald M. Dickey. It is certain that a post office was located along the stage route in the southeast corner of the township. Information from several early interviews leads me to feel quite certain that the next two postmasters were located near the western county line along the stage route.

After Wilton lost the race to become the Waseca County seat, the stage route changed and authorities declared that the post office should be located at Meriden Station, as it was commonly referred to. On June 22, 1870, Frank M. Goodsell was appointed postmaster. It is reputed that the large house, which stood on the knoll a half mile south of the village until it was razed in 2008, was a stage stop. The supporters say that its long hallway with many rooms was similar to old hotels. I was in that house many times in my youth, including overnights, and knew it well. I am aware of other houses in the area with a similar design, and it is possible that a stage ran north to connect with the route to St. Peter, but I have found no evidence of that.

Four generations: Anna Marie Wilker Scholljegerdes (Mrs. John) 1856-1935; Ida Emila Scholljegerdes Drache (Mrs. Max) 1877-1934; Paul A. Drache 1899-1990; Hiram M. Drache, 1924-; taken 1939.

In the spring of 1932, my parents purchased a new suit for me because I was to be in a four generation picture with my father, grandmother, and great-grandmother. It was then that I learned that we were directly related to two of the pioneer families of the Meriden-Lemond area. In doing the research for this project I tracked the relation back to Christopher Henry Wilker. He was born in Germany on February 10, 1821, and in 1842 emigrated, with his family, to Baltimore, Maryland. Later, Christopher, a brother, and two sisters migrated to Guttenberg, Clayton County, Iowa, where Christopher met and married Louise Ribbe. While still in Guttenberg the couple had a set of triplets who died soon after birth. On August 17, 1852, son John H. was born, and in 1854 Conrad Henry arrived.

In early 1856 the family was among a group of Germans who migrated to Meriden and Lemond townships. The trip from Guttenberg took several weeks using covered wagons pulled by oxen. On June 26, 1856, Wilker received a patent to eighty acres in Meriden and an adjacent eighty acres in Lemond Township. The family lived in the covered wagon while land was prepared for a garden, which probably was planted primarily to produce rutabagas, and built a log cabin.

The original Wilker cabin was built on the Meriden part of the farm and on March 20, 1856, Anna Marie, the first white child, was born in the township and also was the first child to be baptized at in what became the St. Paul's Lutheran congregation. On November 14, 1876, she married John Scholljegerdes, and years later the log cabin was moved across the line to Lemond Township, where a new house was erected. On September 1, 1877, she gave birth to Ida Emelia Louise, my paternal grandmother. Anna's brothers, who remained in both Lemond and Meriden townships, were William and Henry.

A 1951 photo of the Christopher C. Wilker's 24 x 28 foot, two room and loft log cabin built in Section 31 and later moved to Section 6 in Lemond. This was the birth place of Anna Marie Wilker, March 20, 1856, the first white child to be born in the township. Note the yoke. Robert Scholljegerdes photo.

Erma Kriesel, a great-granddaughter who wrote this Wilker history, noted that the rutabagas were the most plentiful food that first winter. "Great grandmother would ask her husband, 'Pa what shall I cook today?' He always answered, 'Cook rutabagas, cook rutabagas'." Rutabagas were popular among the Germans and were a basic food in lumber camps. However, potatoes were far more popular and always carried in pioneer wagons because they were the most reliable to produce a crop in the freshly turned sod of the native prairie.

A team of oxen on a walking breaking plow could open the sod at the rate of one-half acre per day. Sometimes the first breaking was done in the fall, and after the virgin sod had rotted over winter, it was backset or crossed plowed the next spring. Then it was tilled before seeding. In some cases the first breaking could be done early in the spring and, if conditions were right, the sod could be turned again and be ready for seeding the first year. In any case, breaking the virgin prairie was a costly, time consuming job.

There were no facilities for grinding the grain in the area so, after the settlers had grown their meager first crop of wheat, they journeyed fifty miles cross-country to a flour mill in Red Wing to have it ground into flour. To make that round trip would have required the better part of two weeks. Erma's account noted that the Wilker boys played with Indian boys whose families lived in tepees in the nearby woods. As a boy I was on that farm often and recall the large woods. She wrote, "One time when the men were gone an older Indian man, reputedly a chief, came to the cabin and asked for bread. Great grandmother gave him some. He returned and she gave him more. He came a third time and she lost her patience and chased him away. Brave women they were."

During the same period the Welk, Radel, and Hoffman families, all natives of Germany, migrated to Honey Creek in southeastern Wisconsin and then to Meriden. The Radel and Hoffman families settled in Deerfield, but the Welks found land in Meriden. On May 18, 1874, John H., the eldest Wilker son, married Amelia Welk. Five daughters died in infancy but two sons, John C. and Albert H., survived. On March 20, 1856, a daughter, Anna Marie was born to Christopher H. and Louisa Ribbe Wilker.

On September 24, 1856, the first marriage of a township resident took place when William F. Drum, of Meriden, married Roxie Henshaw, who lived just across the line in nearby Woodville Township in Waseca County. The wedding was performed by Rev. H. Chapin, of Owatonna, at the home of Austin Vinton, whose farm was in Woodville adjacent to land the Drum family owned. The guests traveled by oxen, which probably traveled at the top speed of one and one-half miles per hour. The first marriage within the bounds of the township was that of Daniel Root (another source stated Pool) to Rebecca Williams, an original settler of the township.

In July 1857, at what is cited as a significant civic event of Steele County, Mr. F. J. Stevens, of Meriden Township, gave the major oration. He was the first state senator representing district sixteen, which included Steele

County. He was also county superintendent of schools for several years but later returned to Massachusetts to enter the banking business. Other early arrivals were Ashley C. Harris, J. O. Wuamett, E. L. Scoville, Joseph Grandprey, Henry Leroy, Robert Stevenson, Fred Walther, J. D. Backus, F. W. Goodsell, Herman Rosenau, Herman Stendel, C. H. Wilker, Anton Schuldt, William Schultz, Henry Abbe, William Mundt, John Drinken, Thomas Andrews, and Lysander House. Many of these names were not found in later plat books because they only came to secure a piece of the public domain in hopes that it would appreciate in value. This was a common practice during the westward movement. The population of the township in 1857 was 193.

The first school classes were held during the summer of 1857 with Dianthy Leroy, the daughter of Henry Leroy, as teacher. In the fall of 1857 the first schoolhouse was erected on the southeast corner of section four. This was the beginning of district fifty-two. About the same time a second building, a log structure, which is one of the few references to logs, was built in section thirty near where the cemetery is located one mile south of St. Paul's church. That eventually became district twenty-one.

By the summer of 1858, three deaths were recorded. The first was that of Edwin House on May 3, 1858. The second was that of Mr. Simmons, a son-in-law of A. M. Fitzsimmons, the first settler in the township. He was killed by lightning while sitting in his house. The third was of Andrew Cook, with no explanation.

F. J. Stevens was the supervisor from Meriden at the first county board of equalization meeting in October 1859. At that meeting it was reported that the township had the second highest number of acres (16,828), which were assessed at $2.80 per acre. By 1861, 19,749 acres were being assessed with the rate set at $3.01 an acre. This indicates that many of the owners were serious about breaking the sod and preparing the land for farming or improving it for the sake of greater resale value.

Most of the homesteaders came with oxen, which was their major source of power. Oxen were not as susceptible to injury or sickness as horses, and they were capable of fending for themselves when there was little or no feed prepared for them. They were also less costly to purchase, and after their working life was over, they were generally fattened and eaten. Initially, oxen were preferred over horses because they were more powerful, and when breaking soil that contained tree roots, their brute force was essential. But their big drawback was that they were so slow. Until the 1850s they were more numerous on farms than horses. Except for plowing and crude tillage of the soil, most field tasks were done by humans. Once machinery, such as hay mowers and self-rake reapers for grain, which needed speed to function properly, was developed, horses replaced oxen. Up to that time it was not practical for a farmer to have more than forty to eighty acres if he did not plan to employ extra help. These two implements enabled farmers to increase their acreage of hay and grain and enabled a family to operate more land.

Three names appear on the Book of Original Entry whose family names are again on record on the 1879 plat books and at a later date. They are Thomas Andrews, David House, and John O. Wuamett. They came as follows:

John Oleson's name is recorded in the Book of Original Entry for April 7, 1856, in section four, as having received title to 160 acres. On May 19, 1856, Thomas Andrews, of Ohio, who was married to Margaret Blong, a native of Ireland, purchased Oleson's 160 acres with a house for $600 from the Oleson who had Americanized his name to Olson. The house was quite substantial, for it was used by the family until the 1920s. The Andrews farm increased to 307 acres in the early 1900s, but family members always had other business and were very involved in civic interests. The original quarter is now the home to the James W. Andrews family and became the first century farm in the township.

James W. and Katie Andrews. James W. Andrews photo.

David House was born in 1821 in New York and in 1844 married Sarah Chapin, also a New Yorker. On April 8, 1857, they received title to their claim in section twenty-seven. By 1879 the family had increased their farm to 210 acres and had a livestock enterprise.

John O. Wuamett, a native of Montreal, Ontario, emigrated to this country on July 24, 1857, and purchased eighty acres in section three. Waumett was one of twenty-eight Meriden pioneers who served in the Civil War. J. O. Backus, Fordyce Brown, Herman Roseanu, and Henry Walters, Sr. are the only others who I was positively able to identify as veterans. On March 30, 1865 Wuamett married Sarah Glover, a native of Wisconsin. Using his veteran's benefits he had expanded the farm and by 1879 had 320 acres. Most of the land in the township was taken prior to the Civil War, so only a few had a chance to use their veterans' benefit for land in the township. In addition, Wuamett had been very involved in civic activities. That farm was still in the family name as of 2009 and was the second farm in the township to be recognized as a century farm.

Joseph Grandprey was born April 22, 1818, in Three Rivers, Quebec. On September 15, 1844, he married Marinda Middaugh, a native of New York. In 1854 they settled in Lemond Township. In 1862 they decided to settle in Meriden but apparently they were misinformed about its availability. Grandprey walked to Winona, not unusual for those days, and learned that someone had filed in 1858, but he was able to purchase eighty acres of that property. The family owned that land until son Samuel E. sold it and moved to Meriden in 1891, where he became one of the developers of the village. The Grandprey name was prominent in Meriden until 1977 when both Lloyd and Eva died when their home was struck by lightning and burned. Lloyd was the town clerk, and because their home was a safer place for the town records than the town hall, which was above the blacksmith shop, the town board felt he should keep the records at his home. Most of the town records were burned in the fire. [ii]

The Railroad–A Key to Development

By the 1850s the western territories and states were very aware that the best way to attract people to the remote areas was to have better transportation, so they pressed the federal government for generous land grants that could be converted to cash to pay for construction purposes. There was little reason to till the soil unless there was a market for its produce. Subsistence farming was no longer in vogue.

At the same time that Iowa and southwestern Wisconsin were being settled, the military, along with fur traders, trappers, and lumberjacks, followed the Mississippi up to where Fort Snelling was established near the Falls of St. Anthony. By 1853, when the first settlers came to Steele County, Minnesota had a population of at least 150,000. The area that would become Steele County was close enough to that population so it did not have the threat from Indians that immigrants only fifty miles to the west experienced.

On March 3, 1857, Congress passed a land grant for Minnesota and made the state the agent to give alternate sections for six miles on either side of the right-of-way, but if some of that land was already taken, the grant could be extended to fifteen miles. The territorial legislature responded with a special session on April 27 by passing legislation that would help attract settlers to Minnesota and beyond. It passed a land grant that approved a route for the Transit Railroad Company to build a road across the state from Winona west to a point south of the 45th parallel to the western border of the state. The Transit Company ran into financial problems, so the state assumed the assets and transferred them to the newly organized Winona & St. Peter Railroad Company (WSPR). However, the Panic of 1857 created financial problems for the governmental bodies and potential investors of railroad stock or land. Before the impact of the Panic was over, the Civil War broke out in 1860. Up to December 3, 1862, the WSPR had been able to complete only two miles of track.

In 1860, Steele County was divided into commissioner districts. Meriden became part of district two, which also included Owatonna and Dover, the original name of Havana. The state again interfered with the life of the people of the township when on February 12, 1867, it deeded 1,234.1 acres out of the 23,040 acres in the township to the railroad. It had forty acres along the siding in sections seventeen and eighteen in the village and 207.34 acres in section five, which contained Bradley Lake and six scattered quarter sections. The railroad grant consisted of the odd numbered sections in a strip that ranged from ten to forty miles wide. No problem was encountered with the existing owners.

The Homestead Act and the railroad land grants that came in the following decades made it possible for waves of homesteaders to acquire land and secure financing as land prices rose steadily. Unfortunately, interest rates were high and many failed, but those who succeeded supplied the nation with an abundant supply of food and fiber. The steady demand of an industrializing nation for agricultural products and the increased productivity of agricultural labor made possible by improved farm machinery caused a steady westward movement.

The WSPR received 1.7 million acres and its successor, the Chicago and Northwestern Railroad (CNWR), 1.1 million acres, most of which was in Minnesota. A total of 130.4 million acres was granted to all railroads. The two railroads mentioned sold their land for from $2 to $10 per acre generally payable in seven installments at 7 percent interest. They also offered a 12.5 percent discount for cash. Both the state and the federal governments wanted the railroads because they were essential in getting people to settle the land and create an economy. The land grants were enacted to provide income from the sale of land, which was to pay for the construction of the roads. Unfortunately, the huge grants presented opportunities for pork barrel politics and fraud. A study of the need for grant reform listed the number one violation as bribery of federal and local officials. Minnesota was one of the top two states in the use of grants with 25 percent of its total area deeded away. James J. Hill profited from one of the early land grants, but he later built the Great Northern with private funds at a cost far less than the land grant roads were built. *iii*

Good farming territory encouraged the railroad to build westward rapidly. By August 28, 1866, it had covered ninety miles and entered Owatonna. In December it had passed through Meriden. By then the railroad operated 10 locomotives, 7 passenger cars, 3 baggage cars, 205 freight cars, and 31 flat cars and maintained an average of 300 employees. In 1866 it exported 3,256,482 bushels of grain and handled 67,000 tons of imports. In October 1867, a total of 103 carloads of grain arrived in Winona, and 44 cars of freight were shipped west on the line as far as Waseca. In 1867, a section crew was established in Meriden and was assigned the section of tracks east to the line between sections thirteen and fourteen where today it crosses two roads, and to the west it was responsible up to the first switch entering Waseca. That was a total distance of about nine miles, which remained the Meriden section's responsibility until 1956.

While the above activity was taking place, the site for Meriden was selected by the railroad surveyors, and in 1866 it was marked on the map as Meriden Station. In the day of horsepower farming, the railroad established a place for an elevator and, if needed, a stockyard every eight miles. The railroad planned for a water tower, a stockyard, and also for the locomotives. Where smaller locomotives were used, water towers had to be spaced from thirty to fifty miles even if there was no stockyard. Waseca, which was already established only six miles to the west, was a major stage stop for a north-south route and west to Mankato and St. Peter. In 1870, Waseca won the county seat contest with Wilton partly because it was on the survey route for a north-south railroad, which helped its cause.

The CNWR erected a small depot on the north side of the tracks and a typical flat storage building to be used as a grain elevator slightly west of the depot on the south side of the tracks. The first account of the elevator stated that E. L. Scoville was the grain buyer. Because grain wagons were not yet in vogue, farmers hauled grain to the elevator in sacks, chiefly in two-bushel canvas Bemis bags. The grain was emptied into bins and later loaded into a rail car. This was done by an elevator driven by a one-horse-power sweep. Scoville came as a single man and purchased an odd parcel of fifty-one acres along the railroad from one of the original owners. His parents, Seymore J. and Mrs. Scoville, came in 1869. Seymore was the son of a veteran of the Revolutionary War, which gives an indication of how new Minnesota was in relation to the nation's history.

By 1867, the road had constructed 105 miles, which included Waseca. In the same month, the WSPR let contracts for extending the line to Janesville. In November 1868, the CNWR took over the Winona railroad because the WSPR did not have the assets to compete with the larger roads in southern Minnesota and northern Iowa. Then, the CNWR planned to extend its line from Madison, Wisconsin, to Winona.

In February 1869, the railroad had two passenger trains each direction going through Meriden. They were both subject to flag stops. The conductor on one of the passenger trains stopped the train west of Meriden to hunt prairie chickens, and one of the passengers was very upset when the train arrived in Waseca too late to make his stage coach stop, which traveled north out of that station. This is proof that travel problems have not changed even though the mode of travel has.

Other Activities

Herman Rosenau enlisted in the Union Army on February 6, 1865, and was fortunate to get an early discharge after peace was declared. He returned to Meriden and in May 1866 acquired lots in the village, and in 1867 presumably used his veterans' benefits to acquire 240 acres two miles southeast of the village. He later opened a saloon in the village and then became a farmer. His son, Henry J., later became a very successful butter maker and then an excellent farmer. Veterans were entitled to 160 acres, and the Preemption Act gave squatters the right to settle public domain before purchasing it. Probably the highest ranking veteran to settle in the township was Dr. Fordyce Brown, "a respected doctor who lived here for many years after serving in the Civil War," according to his mother Mrs. H. M. Brown, who notified the *Owatonna Journal* of his death.

The *Owatonna Journal*, hereafter *Journal*, is available on microfilm starting from 1868, so it is possible to follow what was taking place in the county. After local correspondents were engaged, even the townships news was available. For example: Meriden had four delegates to the county Republican convention because it was allotted one delegate for each twenty-five Republican votes. The Democrats had two delegates based on one delegate for each fifteen Democratic votes.

In April 1868, a notice in the *Journal* listed 145 parcels of land for sale because the taxes for 1867 had not been paid. The property would be sold June 1 if not paid by then.

Paying taxes was made easier by 1869 when the railroad completed larger station houses in Meriden, Havana, Claremont, and Dodge Center. Each year after that the county treasurer placed a notice in the paper informing the taxpayers that he would be at the depot to receive their payments. Because there was no other suitable building in the village, the depot was frequently used for township or social meetings. The poll tax was still in practice, so taxes often were paid by working on needed projects. Farmers often came with teams to do road and ditch work.

In May 1868, the Waseca price per bushel for wheat was $1.65; corn, $1.00; oats, $0.73; potatoes, $0.60. By December the prices for those commodities had dropped to $0.78 for wheat and $0.35 for all other crops. Local prices for grocery products were eight pounds of brown sugar, $1.00; three and one-fourth pounds of coffee, $1; one pound of rice, $0.15; a pound of soap, $0.15; raisins, $0.30; crackers, $0.30; and a three-pound can of oysters, $0.40.

In September 1868, the school lands, which by law were sections sixteen and thirty-six, were for sale. The local officers had decreed that the price would be no less than $5.00 per acre or appraised value. The other domestic news was that Kate Reemsnyder, of Meriden, was married to P. P. Smith by Rev. E. H. Alden, a traveling parson with the German Methodist synod.

The Meriden correspondent had the final comment of the 1860s when she wrote: "There is nothing like a constant stirring of the soil to make the corn grow in dry weather. Corn is very backward for this time of year, but with thorough cultivation it may make a crop. Some . . . have gone through their fields as many as five times and will keep a horse going until harvest." Obviously the farmers were using a one-horse walk-behind cultivator. Contemporary farmers are not likely to agree with those practices. *iv*

---

[1]   James A. Child, *History of Steele and Waseca Counties, Minnesota: An Album of History and Biography* (Chicago: Union Publishing Company, 1887) 21–23, hereafter Child; Hiram M. Drache, *History of U.S. Agriculture* (Danville: Interstate Publishers, Inc, 1996) 71–73, hereafter Drache, *History*.

[2]   Erma Kriesel, "A History of the Christopher Wilker Family." This seems to have been written in 1979 when Erma was about fifty years of age .

³ Child, 32, 39; Hiram M. Drache, *The Challenge of the Prairie* (Fargo: Institute for Regional Studies, 1970) 61,65–66, 140, hereafter Drache, *Challenge*; Child 328–329; Alan R. Woodworth, *The Genesis & Construction of the Winona & St. Peter Railroad, 1858–1873* (Marshall: Society for the Study of Local & Regional History at the History Center, 2000) 1–7, hereafter Woodworth; Minnesota Historical Society, *State Land Office Approved Lists: R.R. Land Grants 1860–1931*: SAM 45, roll 57, book 5, 456; book 6, 105, 122, 127, hereafter MHS microfilm roll 57; Harold F. Peterson, "Early Minnesota Railroads and the Quest for Settlers," *Minnesota History Magazine*, XIII, 21, 25, 29; Daniel J.Elazar, "Federal-State Relations in Minnesota: A Study of Railroad Construction and Development," diss. U of Chicago, 1957, 11, 12, 35–36, 38–39, 45, 60; George Draffen, *Taking Back Our Land: A History of Land Grant Reform* (Seattle: Public Information Network, 1998); Hiram M. Drache, *The Day of the Bonanza* (Fargo: Institute for Regional Studies, 1964), hereafter Drache, Bonanza.

⁴ Woodworth 10; Raymond Evans, section foreman, personal interview, 10 June 2002, hereafter Ray Evans; Lloyd M. Grandprey, personal interview, 11 September 1976, hereafter Grandprey; Child 206, 210, 330; *Owatonna Journal*, 23 April 1886; *Waseca News*, 7 April, 15 July 1868, 25 August 1869, hereafter *News*.

CHAPTER II
The Village is Established
1870–1884

The Chicago & Northwestern

In 1860, when the Civil War started, the nation had 30,000 miles of railroad, which gave the North a decided military advantage by being able to move huge quantities of goods over great distances. After the war this was the key to opening the vast open spaces of the frontier. As previously stated, stations were established about every eight miles, which meant that farmers along the railroad tracks would not have to travel more than four miles to deliver their grain and shop for their needs. In the 1860s, horses replaced the slow moving oxen, so if travel conditions were good, farmers could make a trip to town in about an hour.

But the lack of roads explains the lonesome life of those days. Horses needed to be saved for farm work, and the heavy wagons were not comfortable for riding. This explains why, after roads were constructed, the farmers who could afford them purchased lighter "road horses" and buggies with seats so they could travel at a faster pace. The following is a good example of the effort it took for a farmer in the 1920s in an area where there were no roads to get his grain to market. It took four horses pulling a wagon with a triple box loaded with eighty-five bushels of wheat from 5 a.m. until 7 p.m. to travel the nine miles from his farm to the elevator and back. If he made the trip on a cold winter day, he had to walk along side the wagon to keep warm.

In 1867, after the WSPR arrived in Waseca, passenger traffic from the east improved because Waseca was an important stage town for the north-south stage and to Mankato, St. Peter, and other points in the Minnesota River Valley. On January 16, 1871, the bridge across the Mississippi was finished, giving the Winona road direct access to Chicago, so it extended passenger service to Chicago and Milwaukee. By July 1871, the road had 500 men employed laying track between St. Peter and New Ulm.

Being able to deliver grain directly to Chicago and to Milwaukee caused an immediate drop in the rate from $0.31 to $0.24 a bushel. At that time wheat was $1.00 to $1.06 per bushel, so the reduction in rate was meaningful to the producer. Improved transportation was still costly, but it

Goose Lake snow blockade on the railroad. Minnesota Historical Society (MHS) photo.

helped to open the world markets, which had a positive impact on Midwest agriculture; for it gave our nation much needed exports to fund our industrial growth. For example, No.2 wheat that sold locally for $0.83 was sold for $1.09 at the Midwest terminal and sold for $1.74 a bushel in London. The entire system was not as efficient as today, but it opened the way and everyone gained. The CNWR declared that on its 140 miles of track the cost of hauling a ton of wheat was $4.00 per ton, 12 cents per bushel. Construction continued, and by October 1872 the road was completed to the Dakota boarder.

The future of the CNWR received a boost in March 1872 when railroad magnate Cornelius Vanderbilt took control of the road including the WSPR. Immediately orders were placed for eight new locomotives and 500 boxcars to be added the existing fleet of 200 boxcars and 150 flatcars. Owatonna had shipped 70,037 bushels of wheat east and had received five times more lumber and other freight than any other station on the line. In February 1873, the road had completed more line, and Minnesota deeded 22,666 acres that originally had been provided under the WSPR grant. In May, while a train was backing from the junction at St. Peter to Mankato, it hit a cow on the track and derailed the rear car, which in this case was the front car. Despite the mishap the line from Mankato to Marshall was completed by May, and immigrants were soon moving into that area. At that time Meriden had two passenger trains in each direction daily. This meant that it would not be long before people from Meriden could travel to Chicago and points east without changing railroads. Initially, the passenger traffic was not heavy, but the trains had mail contracts that required the frequent service. At smaller stations that were served only by flag stops passengers sometimes rode in the caboose of the local freights when they made their daily runs.

A news article at that time indicated that freight cost per mile by wagon was 200 mills per ton mile. By rail the rate was thirteen mills per ton mile, and by ocean six and one-half mills, with river barges the lowest at three mills per ton mile. On December 4, 1873, the WSPR interests announced that they had agreed to sell their elevators along the line and from that date on all railroad listings were made as CNWR. As the communities along the line became more populated, and the CNWR finances improved, it added fifteen more passenger cars to keep up with the growing demand. New boxcars thirty-four feet in length with a 40,000-pound capacity became the smallest standard cars. To accommodate the heavier engines and cars, the road replaced the original iron rails with heavier and stronger steel rails on the track from Owatonna to Chicago. Meriden benefitted, for in 1882 the original depot was dismantled and a new larger structure was erected.

A tornado traveled through southern Minnesota on July 14, 1883, and derailed a passenger train consisting of two passenger cars, a mail car, and a baggage car that were thrown off the track about four miles east of the village. The rear coach was the only car that remained upright, but the engine was the only part of the train standing on the track. A yard engine from Owatonna brought the wounded to that city. Thirty-one passengers were injured including Mrs. Blume, Fred and D. Bruggermann, all of Meriden. Rochester had forty seriously wounded and twenty-nine killed. [1]

A 1883 train wreck east of Meriden, caused by a tornado. George Joriman photo.

Plotting the Village

E. L. Scolville lived in the house that was later lived in by Gust Schendel on the south side of the east-west road in Meriden often referred to as Mill Street. (That name will be used for easier identification; sometimes it was called Pig Tail Avenue because of the stockyards.) East of there Fred Fette later built the large white house that still stands south of where Hayes Lucas stood.

In 1868, Mrs. E. E. Goodsell built a house on a two-acre lot on Mill Street for $600. This later became the Paul Drache home and housed the Meriden post office from 1928-1968. Lloyd Grandprey told me that this structure was partly of logs, but my father said that he had never spotted any logs while working with the walls in the sixty-five years that he lived there. I have not been able to determine that any building in Meriden was made with logs. When Frank Goodsell became postmaster in June 1870, a small building from the Goodsell farm located north of where the creamery was built, was moved in and located near the schoolhouse to serve as the post office. Grandprey recalled his parents saying that it had a badly warped floor. The next house located east of the one referred to as the Gust Schendel house was built somewhat in the shape of a schoolhouse by a Mr. Eggert. It was later owned by Fred Eggers, a long-time township resident. The next house east on the south side of Mill Street was an unpainted small, two-story house that was built by W. and Bertha Heinz. Bertha was widowed, and during the 1920s and early 1930s she lived off the charity of the local people. She collected coal from along the tracks and received sawed ties from the section workers for her fuel. I remember her as always being alone on her regular trips around the village until her death in 1936. Even though this was by far the most modest house in the village, families lived there in the 1930s because of the housing shortage in Meriden. There was one more house to the east on the south side. The last residence on the south side of Mill Street was a fifteen-acre farm with a very substantial house, a combination cow, horse, and hog barn with an attached corn crib, a chicken coop, and a machine shed. This originally was owned by Bill Stoltz who had a threshing rig and also worked at the creamery.

On the north side of Mill Street starting from the east was a house owned by Pat Mayer, the section foreman from 1882-1891. That house was later moved one lot to the west and in the 1920s was owned by Aurelius (Rollie) Mueller, long-time section hand.

At that time the Meriden section had reached its full extension, which required at least three people to keep the track properly maintained. All of the remaining property along the track north of Mill Street was in the Goodsell name until into the late 1870s when the Meriden Milling Co. erected a large red building that stood near the rail siding east of where the stockyard was located.

After the creamery was built in the 1890s, the north-south road through Meriden was often referred to as Butter Milk Ave. This was a county road and will be called that for ease of identification. Some time in the 1870s the German Methodist Evangelical Church was built south of the tracks west of the county road. There were other churches of this denomination located in Deerfield, Lemond, Auroa, Bixby, Owatonna, and Waseca. To the north near the tracks was the Central Hotel. The only property north of the tracks to the west of the street belonged to the Goodsell family. On the east side of the county road, L. Peter owned seventy-two acres for farming, and a southeast corner lot against the tracks was occupied by the Meriden Hotel, later owned by Oliver Abernethy.

In 1868, the depot was designated as the voting site. The other legal business for the year was the announcement establishing the tax rate for the three school districts, #10, #21, and #48; no rate was over seven mills.

The *Waseca News* of May 11, 1870, stated that a Mr. McCall, the Meriden blacksmith, had made plans to move west. This was the first reference to a blacksmith in the village. Blacksmiths were very important in farming communities, so it was good news for the village that Peter Pump, a trained blacksmith, moved in and opened a shop. Pump was born in 1845 in Holstein, Germany, and arrived in western Illinois in 1866 where he worked in his trade until he made the decision to relocate in Meriden. Mary Vogelman, a native of Wittenberg, Germany, came to Meriden in the late 1860s, and on September 14, 1870, she and Pump were married. Pump was a successful blacksmith and ran the business until 1875 when he purchased two farms located southwest of the village. In 1898, the Pumps retired and moved to Meriden where they built a large house that still stands. It is the third house north of the gravel road that travels west out of Meriden. Their children who remained in the township were Emma, who became Mrs. Carl Enzenauer; Martha, who married Carl Kujath; and son Theodore, who took over the farm when his parents moved to the village. On May 2, 1870, A. M. Dickey, a partner in the store with E. E. Goodsell, sold his share to Goodsell, who reputedly later partnered with John Wuamett.

The big scandal in town that year occurred when Mr. Hatch was bound over $250 for his assault on R. W. Middaugh for trial in December. The *Journal* reported, "His paramour, the widow, was bound to keep peace for the same amount." Middaugh was hitching his horses to a dray that was en route to Hatch's place. When Middaugh did not move, Hatch beat the

Middaugh horses with a six-foot club and then turned on Middaugh, inflict-
ing serious injuries.

On February 15, 1868, Henry Lutgens signed a mortgage to William
H. Mattson in the amount of $1,033.28 secured by 160 acres. Lutgens was
in default and the property was sold at public auction February 25, 1871.
Lutgens was the only bidder and retained ownership for many years. This
was the first foreclosure notice of Meriden property that was found in the
Journal, which published those proceedings every month. In another case
George and Louisa McDonald mortgaged eighty acres for $225 at 12 per-
cent to Theodore Smith. They defaulted and it was to be sold at auction on
March 23, 1872. None of the four names in the above transactions appear
on any plat, which is a good indication of how fluid the ownership was in
those early years. From the 1880s to the 1950s the turnover of ownership
was very low compared to what took place after the 1950s.

In 1873, Henry Palas, who owned 320 acres of the west half of section twenty along the county road near St. Paul's Church, built a large house
that cost $3,500. The next year he built a 45 x 75 foot barn that had cattle
in the bottom level and a hay storage loft above the livestock. Everyone
who used the road between St. Paul's Church and the village could see that
large house at the end of the quarter-mile driveway with the large barn in
the background. That building site was razed in 2010 when a four-lane
highway was built through the township.

FARM RESIDENCE OF **H.PALAS**, SEC.20, MERIDEN TP. **STEELE CO.,**MINN

A 1874 lithograph of the Henry Palas farm S.W. 1/4 Section 20.

Peter Eliason was born February 10, 1854, in Sweden. In 1879, the Eliasons, along with others from Meraker, Norway, arrived in Meriden Township where the Eliasons settled in Deerfield. That farm later became a century farm. The settlers in that group established the Norwegian Meraker Cemetery in Deerfield, which today serves as a reminder of those pioneers. The Eliasons had ten children who attended District 92, which their father had helped build. In later years, besides his service on the school board, Eliason served on the Meriden Creamery board and the Deerfield Mutual Fire Insurance Company board, which he helped found and served as an agent for many years. He also was a founding member of the Deerfield Rural Telephone Company.

The Deerfield Mutual Fire Insurance Company became an important and durable institution for the farmers, first of Meriden and Deerfield townships and later in Steele County. The Chicago fire caused a great number of insurance companies to fail, which discouraged many people from carrying insurance. At the same time mainline insurance companies sought to discourage people from joining the mutual companies, but the Deerfield and Meriden farmers persisted, and in 1881 formed the Deerfield Mutual in which several Meriden farmers became key leaders. Eliason was a staunch patron of the village businesses. [2]

Changing Agriculture

If someone had lived in Bible times and reappeared on earth in 1830, he could have picked up some tools used back then and gone right to work. But if that person died in 1830 and reappeared in 1890, he would not have recognized a single implement or what to do with it. The new implements invented in the 1830s brought about a revolution in food production. The all-steel plow and the self-rake reaper became a reality, followed by the two-row corn planter and the grain drill. But farmers were slow to adopt what seemed to be costly equipment because they did not realize how much more productive using it would be. They still depended on the local community and relied on the subsistence way of life. They made their own butter and made candles for light out of tallow, a product of home butchering. If they did not use candles they used "car oil," which was made from coal. It was less smoky and smelly than tallow candles, but it still required daily cleaning of the lamp chimney. The kerosene that we use today is a petroleum product, which came later in the century. Grain was cut with a scythe and bound with stalks of grain into sheaves, for they did not have twine or wire. The fortunate ones who could afford a self-rake reaper had to employ two workers to rake the grain from the platform, which then had to be hand bound. Even with that meager equipment, it was estimated that about $500 was needed to establish a farm.

However, it was a period of rapid change in the industry. Clarence Danhof, an agricultural historian on that era, wrote that farmers prided themselves for their independence. He quoted a farmer from that era who observed that there were four classes of farmers: (1) those who were always poor; (2) those who barely made a living throughout their lifetime of farming; (3) those who made a comfortable living and were always improving their competence; and (4) those who became wealthy. It was up to the person to determine to what class he would belong. Fortunately, transportation and a growing urban population were commercializing agriculture, and the industry changed rapidly. This was hard on those who were not willing to change.

The Civil War created a demand for food and manpower that caused prices of farm produce to rise. Farmers could justify purchasing more horse power and the machines that were adapted to their use. This led to freeing the unpaid women and children from their never-ending toil needed to help the farm survive.

In the 1850s, the first livestock was shipped by rail to urban slaughter centers. In 1861 the first "butter plant," i.e. creamery, was erected in New York. Both of these moves were aided by the advent of refrigeration. In the 1830s, the first agricultural fairs were established, and in 1868, the first farmers' institute was held, both of which had as their goal to upgrade farming systems.

Our early leaders were aware that exciting changes were taking place in European agriculture and knew that American farmers had to do the same. Farmers were still the majority of the population, and the myth that farmers were superior to other sectors of society was perpetuated by the politicians because that was the way to win elections. Railroad land grants were enacted in the 1850s, and the Homestead Act and the Morrill Act, both of 1862, were the start of much favorable legislation for farmers. These and similar acts hastened the growth of our nation. As manufacturing and other industries grew and our society urbanized, the influence of the myth of farmers being superior declined. This was a disappointment for many in the agricultural community. It was not until the early 1930s that farmers learned the myth of superiority was no longer in vogue. But in the 1880s the visionaries in agriculture and others understood that agriculture, while still a basic industry, had to change if our other sectors and the nation were to prosper.

F. J. Stevens, one of the township's early settlers and the person who suggested the name Meriden, was on the town board and became one of the first from the township to be on the board of the Steele County Agricultural Association.

There were visionaries and contrarians. For example, in October 1871, a Meriden farmer who signed his column "J" wrote a page-long article for the *Journal* on the virtues of raising sheep, stating that it was as profitable as dairying. However, after the Civil War ended the price of wool dropped because cotton production had bounced back. But "J" contended that sheep were still good. The following week he continued his story on the merits of raising sheep, this time stating that wool and mutton were two great products. Sheep were never popular in the township except for those farmers who used them as lawn mowers around the farmstead. However, they were marketed on a regular basis at the local stockyard.

The scale of agriculture has changed slightly since 1871 if it is gauged by the following account. A Mr. Buffon who spoke at the Steele County Agricultural Association stated that in 1871 he milked sixteen cows that were fed thirty tons of hay he raised on twenty acres and that he supplemented with bran that cost $8.00 a ton. He sold milk to a cheese plant for $425 and received $90 for calves sold. In addition, he earned $132 for 600 pounds of butter sold to a Minneapolis firm, plus all the milk and butter that the family needed.

Changes were already taking place, for the *Journal* reported that a cheese factory had opened in Meriden. The article stated that Berlin, Dodge City (in Merton), Havana, Medford, and Owatonna townships already had plants. I did not find out what happened to these plants, but this was a few years before most of the creameries were founded. See Hanson's *History of the Creameries of Steele County* for that story.

After potatoes, rutabagas, and other garden crops, wheat was the obvious crop to grow on the frontier because it was needed for flour. The agricultural census reported that in 1872 the county farmers produced 287,355 bushels of wheat; 122,140 bushels of corn; 33,987 bushels of barley; and 5,031 bushels of rye. In addition, 150,613 pounds of butter and 2,570 pounds of cheese were produced. The *Journal* editor disputed both the cheese and the butter figures. He wrote that the four cheese plants each produced 2,000 pounds of cheese weekly and listed the number of cows that supplied milk to Owatonna, Dodge City, Havana, and Berlin plants to make butter.

The agricultural base continued to expand. In 1873 the county had 47,170 acres of cultivated land, which included 952 acres of cultivated hay land. At that date about 17 percent of the total area of the county, not the tillable acres, was cultivated. Farms had 3,426 milk cows; 3,094 sheep; 2,003 hogs; 719 oxen as working cattle; and 1,576 horses.

On Thanksgiving Day in 1873, the first activity by any farm organization took place when the Patrons of Husbandry local Rising Sun Grange held a three-hour social in the village hall. The local leader was M. L. Deviny. A three-hour meeting was held, after which the table was set with goose, hen, and turkey. "We retired when the stomach cried aloud for mercy." Someone commented, "They had never had such old fashioned entertainment in the eight years they lived here and it made them feel like they would try for another eight." In January 1874, Samuel Hawks, master of the Meriden Rising Sun Grange, mentioned that their new hall was completed. The A. T. Andreas Historical Atlas of Minnesota, 1874, listed thirty-six farmers in the township as members of the Grange. At that time the Grange had about 475,000 members and was near its peak in political influence.

The other involvement of Meriden farmers in area activity was by J. P. Jackson and H. Warren, who were on the finance committee of the Steele County Anti-Horse Thief Association.

A technological revolution was in progress, and throughout the year there were many ads in the *Journal* about reaper and threshing contests between Marsh, Deering, and Buffalo-Pitts, the most common names. The Buxton and May Hardware of Owatonna advertised that John Deere plows guaranteed to scour. On one page the Woods harvester-self binder, the Adams and French Harvester, (not a self tie), the Marsh Harvester, Masselon Harvester, Minneapolis Harvester, Edwards Harvester, Buckeye Mower and Dropper, and McCormick Harvester were advertised. All except the Buckeye and Woods required three people to operate—poof that farming was changing.

In 1875 the township had the largest wheat acreage in the county—6,239 acre's—that produced 112,900 bushels, followed by 1,243 acres of oats that produced 52,855 bushels, and 1,175 acres of corn that yielded 31,915 bushels. The county had increased to 39,488 acres of wheat that yielded 733,334 bushels, and cow numbers increased to 5,721. The township land was assessed at $12.03 per acre, up from $6 in 1872, which was the highest in the county. In 1878 the assessor reported 461 apples trees. Both of my grandparents had well established apple orchards, which provided fresh fruit for pies, canned fruit, and cider.

James A. Harris was a strong advocate of feeding cattle and learned that Europeans used potatoes for feed, so he tried them and was convinced that when potatoes were priced at $0.25 a bushel it was more profitable to feed them to cows than it was to raise wheat. He sold thirty-three cattle fattened on potatoes and shorts, a by product of wheat milling, for $1,100. His twelve top cattle had brought $600.

Farmers made steady progress, even during the decade that included the Panic of 1873 and a severe outbreak of Blast in 1881, which prevented the wheat from properly maturing. They continued to develop the land, and by 1879 over 22,000 acres were on the tax rolls at a value of $12.73. This produced an assessment of $0.172 an acre, again the highest in the county. John Wuamett told the publisher of the *Journal* that he had a splendid corn crop and fifty hogs and forty-three head of cattle. "He was not troubled about a light wheat crop." [3]

Daily Activities

The primitive state of agriculture mentioned in the previous topic indicates how simple life in the primary industry of the area was and how much time it took just to provide a subsistent life style while the broader society was setting a faster pace. The following news items illustrate how people encountered difficulties, enjoyed themselves, and thrived in that changing society.

In the early years the *Journal* published foreclosures and extensive lists of tax delinquencies. For example: (1) In July 1866, William Suring mortgaged eighty acres to Victor A. Williams and defaulted. A public auction was held February 1872 at the bank at Owatonna. (2) On November 7, 1870, Benjamin Siers mortgaged eighty acres to George W. Newell for $446 and defaulted within one year. The property was sold at a public auction. (3) George and Louise McDonald mortgaged eighty acres for $225 at 12 percent interest on April 15, 1871. Only months later they defaulted and the property was sold to the Bank of Smith Mills. However, it appears that the township had relatively few defaulted mortgages and also had a very good record in making tax payments on time. Frontier life was not easy under any conditions. Some of those who defaulted came in an attempt to make a fast dollar through land appreciation and had delinquent tax accounts because they had no intention to pay. Proportionately, there were more delinquent tax accounts on the lots in the village.

During the years included in this chapter the highest personal property taxes were paid by Anton Fisher, $89 and Thomas Pegg, $33.89, in 1878, and Fisher, $39.80, and Pegg, $94.30, in 1881. The greatest number of delinquencies came in July 1879 on the 1878 taxes when fifty-two individual accounts were listed ranging from $0.77 to $31.39. The highest non-property tax, bill was the $50 county saloon tax which was paid by the hotels and saloons in the village. The most persistent delinquent tax account was the Meriden Mill Co. The county tax collection for the townships for the quarter from March 1 to June 1, 1882, was $39,429.11, of which Meriden had the highest total of $2,954.60. An interesting side light to the tax story is that the second largest investment in Owatonna was the brewery owned by Peter Ganser, which was assessed at $5,000. This limestone structure was on the hill on the east side of Oak Street. After the brewery closed it was partially demolished and remained that way until it was removed in the 1950s. I remembered that every time I rode up the Oak Street hill.

The number of real estate owners reduced from 194 in 1872 to 173 in 1873, which probably came about because some of the smaller holders had sold their property. Most likely they had purchased land for $1.25 and were satisfied to take a quick profit from appreciation. Others spent more time on the frontier. For example, on December 3, 1873, F. J. Stevens, who received title to his land on June 16, 1857, announced that he would be having an auction for his personal property and would be going to New York where he would be employed as a banker. Stevens had built a larger -than-average operation for that day. The auction listed a team of horses, a team of mules, two sets of harnesses, saddle, a pair of oxen, six milk cows, eight heifers, two calves, and two hogs. A complete set of "farm-ing tools"—Marsh harvester, Woods mower, Dayton rake, Loweth & Howe (later Owatonna) seeder, wagon, fanning mill, plows, harrows, and cultiva-tor (The two-horse single-row riding cultivator had just come into use.)—a cutter, 400 bushels of oats, 200 bushels of corn, 50 tons of hay, other provi-sions, and household goods. The terms of sale indicate that Stevens was not financially pressed, for he offered to finance all items over $15 at 10 percent due January 1, 1875, except the Marsh harvester, where half would be due January 1, 1875, and the balance January 1, 1876. Later Stevens announced that he would extend terms on all notes by six months.

Stevens was one of the better known individuals in the township because of his political and civic activities and apparently was comfortable extending the terms. This was during the Panic of 1873, and some of the buyers might have been strapped for cash. At the time of the auction his wife, Lucy, had title to three forty-acre holdings, which they had acquired earlier. Based on average land prices in 1857 and 1873, Stevens gained at least $10 an acre on his sale to Ernst Ruter, who still owned all of that land in 1897 and still owned part of it in 1914. The last reference to Ruter indi-cates that he was buried in St. Paul's Lutheran Cemetery in 1933.

Economic or political conditions apparently did not disturb the people of the township too much judging from the *Journal* article that stated that the Peoples' Party had its county convention in Owatonna on August 22, 1874. The township was allocated five delegates, but no one appeared.

The quarterly tax for the six township schools in 1874 was $773 of which $271.41 was for District 48 in the village. Other tax money received from the county that year was $237.54 for roads, and $126.12 bounty for gophers, foxes, or wolves. By 1882, the quarterly tax had increased to $944.8, reflecting an increase in students. At that time the county had seventy-seven schools. Roads surveyed and built that year ran from the village north to Deerfield. This road ran north between sections seven and eight until it branched off west toward section six and then north to Deerfield. The other branched to the east to between sections four and five then north. Both of those roads became good feeder roads for the village businesses.

C. G. and Naveissa Hersey came to Meriden from Maine and New York respectively and settled in Lemond Township. On September 28, 1864, their daughter Hattie O. was born. The family moved to Meriden Township when Hattie was about ten. By then she was trained to ride Old Kit, a faithful Indian pony, and became her parents "errand boy." She wrote in a short memoir that one day her father was threshing with the horse-power sweep threshing machine and a knuckle on the sweep was wearing out.

A twelve-horse-power sweep powering a threshing machine
that has a chain elevator to carry the straw to a pile. MHS photo.

The cross-country trip took at least two hours. When she arrived in Owatonna the knuckles for the sweep were put in a sack and placed on Old Kit's shoulders. Two hours later she was at the threshing site. She wrote that she was quite proud of that. Another time her mother sent her to go to where her father was threshing, which was about eight miles, to get some money for her mother for a shopping trip to Owatonna, as she had no money. Her mother pinned an envelope inside her coat pocket and sent her off. Her father placed $11 in the envelope and secured it in the pocket. On the return ride Hattie stopped several times to make sure the envelope was still in the pocket. "I was quite relieved when I placed the money in Mother's hands." Hattie added that she liked riding and was happy to run these errands. That way she got out of household chores, which she did not like to do.

In 1883, Hattie was nineteen and got her first teaching job for May and June in District 43 where the Joseph Grandprey family attended. This was about three miles from her home, so she stayed with them. She was paid $50 for the two months and her board and room bill was $15. This gave her $35, which she used to take a course at Pillsbury Academy to earn a second grade certificate, entitling her to teach rural school for three years. On August 28, 1888, she married Samuel Edward Grandprey.

In 1889, Joseph Grandprey bought a large lot north of the tracks on the west side of the road with two houses on it. The first lot had a 198-foot front and was 132 feet deep. The second lot had a sixty-two foot front. These were the second and third houses north of the tracks.

On April 5, 1876, the last couple of the St. Paul's congregation to be married before the new church was Henry Wilker and Doris Abbe. The service was conducted by Rev. John Schulenburg. A reception held at the Henry Albee (not to be confused with Abbe) home was "a grand affair attended by 60 to 70 guests." In the evening the group was serenaded by Prof. J. S. White's "full band, banjoists, vocalists, drummerists, guitarists, bellists and panoists (sic) with their pans and a choursist (sic) of eighteen voices would compare with Mendeloesohins (sic) Club. Their demands were light—only one keg."

Days later the congregation signed a contract with Charles Karnstadt, architect and builder, of Waseca, to build a church for $2,420. The materials were already on the church grounds and work commenced immediately. The church was dedicated on Sunday, October 1, 1876, with morning and afternoon services. The church was a 50 x 34 foot structure that could seat 400 and had a slightly ornamental steeple that was 72 feet high.

January 1878, the German Methodist Evangelical Church congregation held a meeting at the home of Henry Palas. The same issue of the *Journal* also contained an article about a revival meeting at the Pegg school District 52. It is not clear if this was a religious revival or if it was related to a school affair. The school board and twelve or fifteen others met with the male teacher and advised him not to be "bothered by trouble beyond his jurisdiction."

Based on this October 1879 news item, others had more than school problems: "Frederick Mueller ran away from Meriden to escape from marrying Caroline Mudeking, according to promise. He seduced the girl while living at her father's house." Later in the month, N. G. Seeley entered a complaint against Frederick Daltner and August Daltner for assault and battery. Seeley had acquired a farm on which the Daltners had previously lived. When they returned to the farm to get some items they had left there they got into a hassle with Seeley. A twelve-member jury gave a verdict of not guilty.

The new decade opened with happier events. Three women from the village gave a New Years party for fifty guests. Messrs Wheelock and Domy of Woodville furnished the music, and when the dancing was over a midnight lunch was served.

Oliver Abernethy came to the township in 1869 and owned 220 acres north of the village. In 1879, he became the owner of the Meriden House and furnished "an excellent supper and a room for more waltzing." Abernethy held a liquor license, but the article made no comment if any liquor was consumed at the party.

To break the monotony of February weather, the members of the Crane Creek Literary Society visited the Deerfield Lyceum, which met every Saturday evening, and took part in the scheduled debate. That same weekend the Meriden Lyceum, which had about thirty members from District 52, held a social at the John Wuamett home. The program consisted of literary exercises.

When George and Barbara Hubbard, operators of the Central Hotel from 1879 to 1884 (the building west of the county road and south of the tracks), took a new departure in their business by installing a bar, "some of the first Meridenites who visited became over enthused." On March 1, 1880, "The Hubbards held a grand opening and served free drinks which led to a fight in which between eight or ten participated. Two of the fighters lost their shirts, a stove was demolished, and about 125 slight contusions of the skin were received." Barbara Hubbard was a sister of Mrs. Andrew Matthes and Henry Walther both long-time residents of the township.

Oliver Abernethy, the village justice for twelve years, reported the following case: Jacob Krager charged that Frederich Naills had threatened his life. Naills was arrested but then Krager could not be found. Naills was released to return the next morning, but still no Krager. Then Naills charged

Krager with hitting his wife under the chin with his fist. Krager was caught, arrested, and found guilty. He was charged for the cost of both suits. Both were required to post $200 bonds to keep the peace.

Soon after the above incident occurred the depot safe was blown open and about $75 to $80 was removed. The funds were property of Van Dusen Co., a Minneapolis grain firm. Their agent, Mr. Soule, was in the village buying wheat. The thieves had searched the mill but found no booty there.

When the Minnesota Central Baptist Association met in Owatonna in June 1871, Meriden was represented by Rev. J. C. Weedon, W. Stebbins, and J. P. Jackson. Weedon was also listed as a delegate from Berlin Township.

For some reason the *Journal* did not always report news about the German Methodist Evangelical Church, but the following items prove that it functioned for several years. On February 23, 1881, the Rev. H. E. Young preformed the marriage of Bernard Mathwig and Ida Kuchenbecker, daughter of the E. Kuckenbechers. After the ceremony, about eighty guests were entertained at the residence, "with a great variety of eatables and drinkables, and good music, the company spent a day and night never to be forgotten." On February 8, 1883, Rev. Dewart conducted the funeral service at the German Methodist Evangelical Church of Robert Crosby, seventy-three, who reportedly was a resident of the county for eighteen years.

When the "Dakota Tornado" followed the course of the Winona road from Kasota to Rochester in July 1883, a large number of homes and farm buildings were destroyed. The German Methodist Church "was completely demolished as if it had been under a short range battery directed by Satan. The Meriden flouring mill was unroofed and $2,000 damage done; the depot was moved six inches but stands as firm as it did on the original foundation. Wm. Engles' blacksmith shop was blown to the next lot."

E. S. Wheelock, who was a partner in the Meriden milling firm, announced that he was leaving Meriden for Kampeska, a new town in Dakota Territory. In October the machinery from the damaged mill was sold to parties in Kampeska. By August 31, the *Journal* reported that the church frame work was already up and would soon be enclosed. The Evangelical Association had services on Friday evening December 7, twice on Saturday, and dedication on December 9.

Miscellaneous happening during this period: In 1877, Steele County voted to erect a jail and issued bonds for $10,000 bearing 9 percent interest payable in one and two years. That building served the county for over a century. In 1881, the population of Owatonna had grown enough that redis-

tricting of commissioner districts was needed. Meriden was put in District
1 with Berlin, Lemond, and Summit townships. The first meeting of the
new county commission was a call for a vote for a court house and to sub-
mit bonds for $35,000. A November 1881 article stated, "The recent rains
have proven the need for gravel roads so farmers could bring their produce
to market when the roads are wet." An indication of how critical the lack of
roads was is illustrated by an item of December 1883, which reported that
because of the scarcity of feed, nearly all hogs in the township have been
sold. On the national level, the two-cent postage stamp for first class mail
became effective October 1, 1883. News from Owatonna indicated that
when thirty subscribers signed up the city would be able to have telephone
service. This is an example how technology spread, for Bell had invented
the telephone on March 7, 1876. [4]

---

[1] Hiram M. Drache, *Beyond the Furrow* (Fargo: Institute for Regional Studies, 1976) 214,
hereafter Drache, *Furrow; Owatonna Journal*, 15, 22 December 1870, 19 January, 9 February, 20 July 1871, 11 January, 1 February 1872, 27 February 1873, 20 September 1877,
28 January, 27 February 1880, 1 December 1881, 6 January, 28 March 1882, 22 July 1883,
hereafter *Journal*.

[2] Grandprey; *Journal* 3 September 1868, 23 February 1870, 29 June 1871, 9 July 1874,
1 December 1882, 9 April 1897; Joanie Mosher, "A Brief History of Peter Pump" (Owatonna, 2003 October); Weldon Beese, letter, 22 April 2002.

[3] Clarence H. Danhof, *Change in Agriculture: The Northern United States, 1820–1870*
(Cambridge: Harvard University Press, 1969); *Land of Plenty* (Chicago: Farm Equipment
Institute, 1959) 58–59, hereafter *Land of Plenty*; Drache, *History* 69–76, 317–318; *Jour-
nal* 18 May, 13 July 1871, 21 March, 3 October 1872, 27 February, 1 May, 13 November, 4
December 1873, 22 January 1874, 24 June 1875, 20 July 1876, 1 August 1878, 12 Decem-
ber 1879; A. T. Andreas, *Illustrated Historical Atlas of the State of Minnesota* (Chicago:
Author, 1874).

[4] *Journal* 7 December 1871, 1 February, 25 April 1872, 16 January, 24 April, 20 November 1873, 23 April, 6 August 1874, 14 January, 15 April, 1875, 13 April, 3 October, 1876,
24 January, 14 November 1878, 18 April, 24 October, 28 November, 1879, 13, 20 February, 5 March 1880, 14 January, 4 March, 1 April, 3 June, 11 November, 30 December 1881,
30 June 1882, 16 February, 27 July, 3, 31 August, 5, 12 October, 7 December 1883, 25
April 1884; Child 32, 39.

CHAPTER III
Developing a Farm Service Center
1885–1899

The Chicago & Northwestern

The settlers of Meriden Township were fortunate that a railroad ran through its midst and railroad technology improved rapidly. In 1872, the Westinghouse air brake system was developed, which increased train safety, and the CNWR, which received adequate financing, expanded rapidly during this period. In May 1885, the road put a fast livestock train in operation between Waseca and Chicago. The total time between the two points was twenty-seven hours, which was reputed to be a record speed. Traditionally, wheat was the chief agricultural export from a frontier area because it was durable, so travel time was not critical, and it had a relatively high value per unit of weight, so farmers raised it to have a cash income. With better transportation to improve marketing other crops and livestock, farmers could diversify their operation and improve their diet. In this area chickens, hogs, and dairy cattle were the obvious enterprises to develop. Most of the land was in production, so more grain was being sold. In 1890, the old flat grain storage was removed and a new elevator was built. E. L. Scoville continued as manager.

By then the CNWR had morning, midday, and evening passenger trains going in each direction daily. This made it possible for students to attend high school or work in Owatonna and still live at home. Several were employed at the State School and by 1898, with more students attending OHS, Pillsbury Academy, or Canfield Business College, the road issued commutation tickets between Meriden and Owatonna "for those who needed to travel on a regular basis."

At the same time two freight, trains traveled in each direction daily except Sunday. In addition, a way freight, which picked up less-than-carload lots, and a stock freight ran east only on a daily basis. The WSPR had expanded to 446 miles by the time it became a part of the CNWR, which was part of a 4,210 mile system.

The communities along the line into South Dakota were expanding rapidly, and in the late fall and early winter of 1895 the road ran as many as twenty-one trains to get livestock and grain to market and deliver building materials for the new settlements. During those years an average train consisted of fourteen cars with the maximum load per car of 45,000 pounds. In 1896, to enable the use of bigger engines and larger cars, the road bed was upgraded again, and heavier steel rails were laid replacing the original iron rails.

As soon as the main line was upgraded, the company enhanced its facilities in Meriden. In 1896, a 100-foot well was dug at the stockyard to accommodate the increase in the livestock business. In 1899, the depot was given a new coat of paint. The siding was rebuilt and lengthened to 1,200 feet. A new, larger hand car was provided for the section crew, and a coal shed was built to store coal for use in the depot.

The following are examples of what access to the railroad meant for the local people: In September 18, 1891, C. M. Hanson, owner of the general store in Lemond, came to Meriden to pick up freight. Hanson had to drive a team of horses eight miles to Meriden and return. At the very least it took him two hours each way to make that trip. Hanson had the disadvantage of doing business in an "inland town." The same paper reported that Dr. Hatch of Owatonna made a house call at the Pat Maher home to attend their ill daughter. Dr. Swartwood of Waseca was summoned when Mrs. Louis Abbe was giving birth to a son. The easy access to the train and the nearness of the Mayo Clinic at Rochester made it possible for many people to go there for medical treatment. During the period covered in this chapter one individual went to have his tongue removed because it was infected with cancer and another went to have a leg amputated as a result of a poison ivy infection.

On May, 20, 1892, C. Yust, of Deerfield, was not near a railroad so he had to walk the entire seven miles to Owatonna, "the mud being too deep for successful teamings." In July 1895, Miss Gordon, a teacher of instrument music in Owatonna, was able to take the morning train to Meriden to teach her group in the village and return home on the afternoon train. In February 1896, John Mattice took the train to South Bend, Indiana, to seek work at the Studebaker Wagon Works but was unable to find a job that paid more than $0.60 a day so he returned to the farm. In November 1894, John Scholljegerdes left for Minneapolis on the 6 a.m. train and returned on the 8 p.m. one. On Memorial Day in 1898, a group of twenty-four took the train to Owatonna and returned in the evening. Later that year another group of twenty-five took advantage of excursion rates to attend the Sons of Herman celebration in New Ulm.

Sometimes there was disappointment. The May 13, 1898, *Journal* reported: "A large crowd of Meriden people planned to take the excursion train to St. Paul Sunday but it was full when it arrived and only one person was permitted to get aboard." The lucky person was J. Kirk, an employee of one of the local farmers.

Frank Waumett received $450 and Clara Waumett $237.50 as a result of injuries received while riding a CNWR train on New Years morning in 1892. The company paid out a total of $2,000 because of that accident, which happened just west of Owatonna. In another incident Wm. Dietz, an eighty-year-old farmer and shoe maker, was driving his cow from pasture. "When he attempted to get the cow from getting struck by the train he was struck and killed by the 5:52 CNW train. The train did not stop at Meriden because it was going 'at full speed' County Coroner-Dr. J. H. Adair felt no need for inquest." *1*

Agriculture & Related Activities

Crane Creek

Bradley Lake and Crane Creek were the two largest water bodies in the township. The lake was named after Bradley, one of the early settlers in that area of the township. There is an early account of goods being transported between Owatonna and Waseca by boat via Crane Creek. Crane Creek continued west into Woodville Township then into Watkins Lake, which nearly touched Rice Lake and was about a mile from Clear Lake, but it continued south to Goose Lake. This was all low-lying area that needed drainage. In October 1895, the *Journal* made its first reference to Crane Creek when it was reported that dynamite was used with success in ditching to improve the flow. A large bed of boulders was removed, which became important to draining in several sections of the township. In 1887, Senator Nord of Waseca County introduced a bill in the state senate to appropriate $15,000 to deepen and widen Crane Creek. The farmers along Crane Creek were all well established by the late 1870s and were able to exert enough pressure to secure governmental help. In April 1888, heavy rains caused the bridge over the creek to wash out on the county road going north between sections three and four.

A Changing Agriculture

Technological changes and legislation had a continuing impact on the industry that brought about more change to rural society in the century and one half covered by this story than in all previous history. In 1886, Caroline Woker foreclosed on Henry Abbe, Jr., and wife Augusta because they had defaulted on a mortgage dated 1882 in the amount of $447.56 secured by farm land. To avoid foreclosure the Abbes sold the property for $1,700. The mortgage was $2.90 per acre and they realized $11.04 in their sale, which enabled them to leave with a clean slate and cash.

Many farmers and railroads expanded rapidly during this period, counting on land appreciation rather than good management to make a profit. When the free-land frontier came to an end in the nation during the late 1800s, it caused widespread concern. Prospective farmers lamented that land was priced too high to make a living, but, as always, management-oriented farmers figured out ways to make a profit and, except for sporadic down periods, land prices kept rising steadily. As farming became more capital intense, improved farm management changed farming from a way-of-life to a business-oriented enterprise.

In 1885, the *Hoard's Dairyman* was founded and remained the premier dairy magazine for many decades. In 1887, laws were passed to regulate commerce, particularly railroad rates, and the Hatch Experiment Station Act provided for state experiment stations. The costly practice of driving cattle to distant markets came to an end after 1888 when the first meat was shipped in refrigerated cars cooled by mechanical refrigeration. Agriculture was given enhanced recognition the following year when the USDA was raised to cabinet status.

Harvesting hay with a scythe and pitch fork required thirty-five hours per ton, but by the end of the twentieth century haying could be done completely by machines. Using horses to pull a gang plow, seeder, harrow, binder, wagons, and steam-powered-threshers reduced the time required to produce an acre of twenty-bushel wheat to eight to ten hours. In 1889, the Abbe brothers purchased a horse-power sweep to power their thresher. Owatonna seeders in sizes from seven to sixteen feet were advertised in the Journal.

By 1890, the national population had increased to 62.9 million, of which 29 million were farmers, who made up 43 percent of the labor force. The flow of immigrants to the land frontier kept expanding agricultural production, which proved a boon to the growing industrial sectors.

In 1887, the county paid out $2,314.90 in bounty for the 23,149 gophers that had been trapped and also paid $7 to $8 per wolf that had been killed. Christ Nelson was the township's leading wolf trapper and collected $56 for his efforts. That was a sizeable sum at a time when hired help received $200 to $220 plus room and board for the work season.

On June 24, 1887, my grandfather, Max August Drache, arrived in the United States; he was sixteen. He traveled to Waseca and then walked east into the Meriden area. He remembered stopping at a Wm. Heuer farm and experienced a stay in their log cabin. He located a job at the Henry Grunz farm where he worked for his first year. He only went to Meriden once that year and spent about $10 for clothes. His goal was to save money to return to Prussia and bring his family back.

In 1889, James A. Harris, whose family was among the original settlers in the township, was a member of the advisory committee of the Agricultural and Industrial Association of Steele County. Its primary responsibility was the first annual fair, which was held September 17-19. The next year Lysander House was put on that committee, making them some of the earliest township residents to serve on any county board.

In 1890, Henry Palas, who farmed 398 acres, was referred to by the *Journal* as "one of the oldest and best farmers in the township." He purchased a Nichols & Shepherd vibrator threshing outfit and would be taking orders for threshing. A later item reported that he had shipped two carloads of 107 spring pigs to St. Paul for which he received the top price of $4.80 per cwt. These pigs probably weighed 300 pounds, the popular weight at that time, because lard was in great demand. The So. St. Paul stockyard was erected in the late 1880s, and by this time it received several thousand head of hogs, cattle, sheep, and calves daily. By contrast, the Chicago yard received as many as 75,000 head a day via up to 100 trains. Large stockyards such as Chicago and So. St. Paul, adjoined by packing plants, were the standard means of marketing and processing livestock after railroads came into existence.

The threshing machine was constantly being upgraded. In 1893, Henry Lutgens purchased a Nichols & Shepherd rig with "a stacker attached for straw, one of the first in the county." The so-called stacker was an endless chain conveyor that elevated the straw away from the thresher. Up to that time the straw had to be pulled away from the thresher and often was used by the steam engine.

During the late 1880s, the steel-towered windmill became popular. John Palas who, in addition to farming, owned and operated a steam-powered corn husker/shredder, also sold and erected the new steel windmills. Windmills were a real boon because of the great amount of labor that was required to provide the proper amount of water for the animals. This was especially essential for good milk production and no doubt was partially responsible for poor performance of dairy animals at that time.

Sometimes technology caused repercussions. For example, in July 1893, William Utech went to his cowyard to find six milk cows dead. They had broken out of the cowyard during the night and gotten into the potato field where they ate the potato vines that had been sprayed with Paris green in the previous days. Utech was not the only one to suffer a mishap, for in 1899 many had to dispose of their cattle because heavy rains ruined the pastures and damaged the hay.

Most farmers in the township raised potatoes for domestic use, but those with adequate help raised potatoes to market. E. L. Scoville advertised to purchase 2,000 bushels (120,000 lbs.) of potatoes and most years shipped at least a carload. The two grocery stores often dealt in bulk potatoes and onions.

C. H. Wilker, who came to the township in 1856, turned the farm management over to his son in 1876 and moved to Owatonna. In 1883, he sold his home on Oak Street and retired in California. Wilker had to go to a warmer climate for health reasons and, fortunately, he had the finances to afford doing so. Very few farmers went to California for retirement, but most who retired from the farm went to Owatonna or Waseca. Originally, very few came to Meriden, but that changed in later days.

Early Creameries

Supply and demand took its normal course in the development of the township. As mentioned previously, as population increased and transportation improved, many farmers strove to enhance their income by feeding their crops to livestock   Nearly every farm had a cow or two to supply the family needs and often to have butter to trade for groceries.

Two events took place in the previous decade that greatly aided the dairy industry. In 1879, the DeLaval centrifugal cream separator was invented, and in 1880 evaporated milk was developed. Up to June 1891, the nearest outlet for cream was at the Blooming Grove Creamery, which was about six miles from the village. L. G. Wolter, who was a partner in one of Meriden's stores and had several relatives farming in Meriden Township, was also involved with that creamery. It was a very primitive operation where the cream was skimmed off the top of the milk. The price was $0.10 an inch, and in June 1891, it paid farmers $600 for their cream. Some Meriden farmers advocated that a creamery should be built so they could keep their skim milk at home for calf and hog feed. They felt that it was more profitable to sell butter to Wolter, a popular merchant, than it was to sell him cream.

On August 7, 1891, the *Journal* stated: "Our Meriden correspondent thinks the town will have a creamery soon." The next issue of the paper reported that the creamery was a settled fact. E. J. White, an experienced butter maker, had purchased material for a building to be erected on land adjacent to the village that he had purchased from F. W. Goodsell. White and his partner, a Mr. Lyon, apparently had the creamery in operation by October.

The first creamery built by E. J. White in 1891.

In January 1892, there was news that farmers in the Crane Creek area were planning on building a creamery. Soon work started on the Crane Creek Creamery Co. building on section two. G. Bossard was president of the organization, August Hoffman treasurer, and J. A. Harris secretary. The group moved rapidly and, under the leadership of butter maker Willis Noyes, started operation on April 11 when 2,000 pounds of milk were delivered. By the end of the second week, daily receipts were up to 5,000 pounds. The *Journal* editor was proud to announce that he had received the first pound of butter produced by the Golden Rule Creamery, as the association was called.

In March 1893, board members of the newly formed Crown Creamery in section twenty-five met in Owatonna to purchase equipment for their organization. The equipment arrived in May and soon after Crown was in operation.

Initially, cream separators were costly and not economically efficient for small-scale farms, but they were a boon for creameries. Prior to establishing creameries in many areas, skimming stations were erected where farmers could take their milk to be separated and sell the cream but retain the skim milk for use on the farm. These stations were erected by creameries as a means of collecting cream from more distant farmers. The Meriden skimming station was called "The Big Six Creamery." It was located in section twenty-nine near where the cemetery is located. It was owned by the Golden Rule Creamery, which also had another station at the James Brady farm in Deerfield. It is purely conjecture on my part, but from my knowledge of the farmers in the immediate area of the Meriden station, the "Big Six" name came from six farmers near that station who had larger farming operations and wanted to increase their dairy herds. In 1894, they purchased a Babcock tester, which was perfected in 1890, so they could test the cream content of their milk. This gave them a better idea of what their herd was producing.

The second creamery. It was built by the Meriden Cooperative Creamery Association 1892.
The Clara Goodsell farmstead is in the background. Mary Stendel photo.

The public showed its preference for creamery produced butter. By 1892, the published price for creamery butter was $0.29 and dairy produced was $0.14. I am aware of the fact that grocers who took butter in for trade sometimes molded and packaged it so that customers did not know the source. Grocers gave the farmers more than the cash price for their butter if they traded it for groceries.

The dairy industry grew rapidly, and by 1894, Steele County had eighteen creameries and five skimming stations. Meriden Township, with three creameries and a skimming station on section twenty-nine, was setting the pace for the county. Willis Noyes was rehired for his third year by Golden Rule Creamery at a salary of $900 plus a house; the creamery paid out $25,135 for cream at an average of $0.2387 a pound. H. J. Stebbins purchased the buttermilk for the year for $180, which he fed to his livestock. (For more information about creameries of the county read Bernie Hanson's history.)

E. J. White operated the village creamery until 1893 when he began to experience financial problems, which gave local farmers an incentive to take over. In May 1894, the Meriden Creamery Association was organized with a capital of $2,500. The first stockholders included a good cross-section of farmers from Blooming Grove, Deerfield, and Meriden townships. White remained as butter maker at a salary of $65 per month, and Henry J. Rosenau, son of one of Meriden's pioneers, was employed as an assistant.

In 1895, Rosenau became one of the first to step out of the bounds of the township to enhance his education. He took leave from the creamery to take the butter makers' course at the University of Minnesota St. Paul Campus. After he graduated he returned to become the butter maker. He was an excellent choice for the association. He put the creamery on a more business-like basis and upgraded the equipment. Two new separators and an improved churn increased the amount of cream converted to butter. The churn improved the consistency of the end product. Farmers converted to using cream separators because they realized that by doing so they only had to haul about five percent of the volume to the creamery and had the rest to feed to calves or pigs. That also reduced the volume that the creamery had to handle. By 1896, the creamery was receiving about 42,000 pounds of cream weekly. Thirteen pounds of butterfat yields a pound of butter ,which meant that the creamery produced about 3,200 pounds of butter.

In February 1897, the fifth annual National Butter Makers Association meeting was held in Owatonna. By then the county had 30 members out of 373 national members, of whom 117 submitted tubes of butter that were displayed in the gymnasium at Pillsbury Academy and entered in the national contest. Henry Rosenau won that event with a score of 99 for which he received a silver cup. In July 1899, James Harris, of the Golden Rule association, was employed by the National Dairy Union to work on a campaign to have a tax placed on oleo margarine, which was effectively competing with butter.

In 1898, the village creamery erected a board fence on the west side of the creamery to prevent the teams and wagons from falling off the steep embankment. The association also purchased the Golden Rule skimming station. Louis Palas moved the 14 x 16 foot building to the west side of the creamery where it was used for an ice house. By then the association had ninety patrons and many others who wanted to join. So, in February 1899, it issued $5,000 of new stock. John Scholljegerdes was elected president; A. G. Parke, vice president; Peter Moe, treasurer; and James W. Andrews, secretary. These farmers came from Lemond, Blooming Grove, Deerfield, and Meriden Township respectively. In June a carload of binder twine arrived from the state prison. James Andrews and Peter Moe were responsible for distributing the twine to the farmers who ordered it.

Hammel Bros., of Owatonna, were contracted to design and construct the creamery. The foundation was made of Kasota stone at a cost of $400. Because the local sand pit was exhausted, the members had to form a hauling bee and haul gravel from Owatonna. Hauling such a volume of gravel by wagon teams in December must have been a challenge. As soon as the material was on hand, work began on the 30 x 36 foot building with double walls for the butter making room, plus a 23 x 30 foot addition for the engine and boiler room, and a room for the Babcock tester. A refrigerated room large enough for 140 tubes of butter could hold the temperature at forty-five degrees on the hottest days and only needed to be iced every fourteen days. A fifty-foot tall chimney was built on a separate foundation to complete the construction. The $8,000 investment contained the latest equipment available, and the building became a landmark for the community.

The creamery after the 1899 addition.

Farm Service Businesses

After farming during 1889 and 1890, Sam and Hattie Grandprey decided to quit farming so Sam's parents, Joseph and Marinda, could sell the cattle and machinery, pay off their debt, and retire. In April 1891, Sam and Hattie moved to Meriden where they had already purchased several lots, some with houses. Sam and Joseph enlarged the houses, and Sam also worked at day jobs for local farmers. He became acquainted with N. G. Pratt, a blacksmith and a good wood worker who had a shop north of the tracks east of the county road across the street from the Grandpreys. Hattie wrote, "Pratt suggested that Sam could learn the trade if he would buy half of the tools and furnish the shop supplies and receive half of the income." Joseph still owned a team of horses, and Sam had sufficient funds, so he traveled to Waseca to buy horseshoe nails, coal, and horseshoes for winter needs. That proved to be fortunate, "for it was an open winter with little snow but plenty of ice so there was a great need for horseshoeing throughout the season. When spring came Pratt backed out on his offer. Sam rented a nearby shop and moved his tools to it which left Pratt short of tools so he returned to Waterville." Hattie continued, "In 1892 Sam became the town blacksmith by default but he had the ability to keep the business growing until he secured the help of a professionally trained blacksmith to fine tune his skills. He was fortunate for after the creamery was enlarged, business activity increased considerably and he was able to make a good living." Sometime prior to 1897, he built a blacksmith shop and had the foresight to erect a two-story building, which became the place for many community events in the years ahead. It was popularly called Grandprey Hall.

My maternal grandfather, Wilhelm R. Schultz, came to the United States from Prussia in 1883 and worked in Owatonna and on farms in Lemond Township. In December 1887, he purchased a forty-acre farm from a widow and an adjacent sixty-nine acre farm in Lemond and later purchased forty acres of woodland that adjoined his property. On November 21, 1889, he married Louise Middelstadt, of Owatonna, and soon after they moved to the farm. By then he had built a modest 24 x 30 foot one and one-half story house. The main north supporting wall about four feet high was made of rammed earth. In his native country he learned the process of filling forms with earth and packing it so effectively that it was as strong and durable as a cement wall. When that house was razed in recent years to build a new home for a grandson, none of the people on the project had any idea what it was made of, but it seemed nearly indestructible. The supporting beam that ran the full width was a hand-hewed log. The European practice of using wardrobes instead of closets was used so there were no closets for the one downstairs bedroom and the two upstairs. The upstairs had a small crawl space for storage. In addition to the bedroom, the downstairs had a large living room and a kitchen that ran the full length of the house. One entrance to the kitchen also led directly to a very steep stairway to the cellar, which had a dirt floor. This was home to the family of seven. I never asked, but my mother must have slept on the couch in the living room. My grandparents lived in that house for forty-four years after which they moved to Meriden to live in a house they had built about 1921. Fortunately, the second generation consisted of a very accommodating son and daughter-in-law and their two sons. That farm was never enlarged, but as the son grew older he became involved with drainage ditching and road equipment which proved to be very profitable.

W. R. Schultz steam powered ditching machine on a ditch in sections three and four in Lemond Township on the Schultz farm. This ditch flows to the Straight River.

In 1889, J. R. Petrich and A. J. Speckeen built a 20 x 70 two-story store building with housing on the second floor. The *Journal* stated, "That building will give them more room and conveniences to wait on their numerous customers." It is not clear whether the Central Hotel and store that sat on this site was razed or remodeled and enlarged. However, this building became the center of the community, for the second floor served as a hotel for many years. In later days it often had several families living there because of the housing shortage. This store was south of the tracks west of the county road. Probably no building in the village had more people live in it than this one. Except for the creamery, there was little activity development-wise because of the downturn in the national economy. This building was destroyed by fire in 2003.

Between 1893 and 1894, the volume at the creamery grew by 50 percent, and the village businesses all gained because of that activity. L. G. Wolter, who operated the store, prepared for the winter by ordering a large invoice of boots and shoes, plus he "had plenty of herring on hand." Every week the paper reported that Wolter had shipped thirty to thirty-five cases of eggs, thirty dozen each, mostly to the New York market. John Korupp and his sons were kept busy building barns and houses, including a contract for a Lutheran parsonage. E. L. Scoville reported shipping three or four carloads of livestock each week. But it was the summer drought that kept Sam Grandprey's two employees busy sharpening plow lays because the dry soil was so hard on them.

In 1893, my grandfather, Max Drache, returned to Prussia and convinced his widowed father Karl, his sister Emma, and his brothers Carl, Ernest, and Rudolph to return to America with him. They settled in Meriden except for Rudolph, who settled in Lemond. Max and his sister Emma were both employed by John H. Wilker, a widower. Later, Wilker and Emma were married. In 1895, Max purchased 160 acres for $4,500. On July 14, 1897, he married Ida Emila Scholljegerdes, daughter of John and Anna, from Lemond Township. I have not been able to determine how either of my grandparents acquired the funds to purchase their farms, but I suspect John Wilker may have helped his brother-in-law or else his father-in-law did. I have no idea how W. R. Schultz acquired his land. Max Drache erected a hog barn soon after he purchased the farm, and family lore says they lived there while the house was being built. In 1901, a substantial house with four very large rooms on each floor was erected, plus a lean-to that served as the entrance in addition to providing space for washing the cream separator and a washing machine. The main part of the house had a cemented basement.

A cement block building was erected for the cream separator. Water from the windmill was piped into a cement tank and then flowed to the outside stock tank. The milk house tank always had home brew and homemade root beer in it. Both of my grandparents built a substantial chicken coop, a dairy barn, and a hog house and ran well-diversified farms. The one difference was the family dwelling, which I have described.

In 1895, the national economy improved, and a revival, which lasted for several years, reflected itself in the village. In February 1895, C.A. Litchfield of Albert Lea visited the village prospecting for sites to establish a lumberyard. In March, Litchfield announced that he would open a yard and erect a house for his family. He added that the yard would be a well-equipped lumber establishment. In March, several carloads of lumber as well as cars of lime and brick and a shipment of paint arrived to create the promised inventory. Next, Scoville announced that he had sold his coal warehouse to Litchfield.

In March 1895, E. E. Hocking purchased a lot from F. W. Goodsell just east of the big white structure that still stands (2011), which originally was Peters & Waumett store. Hocking built a house and also a two-story building that held the blacksmith shop with a hall above for public purposes. The building was dedicated with a dance in which the Robinson Bros., of Owatonna, furnished the music. The shop was opened April 1 and soon wagon parts arrived. Hocking manufactured wagons to order and also was a dealer for the Rodecker Roller Window Screen, which was claimed to adjust to any window. He completed his expansion when he obtained a dealership for the McCormick line.

In January 1896, Korupp finished a house for his family at the north end of town and later built a small, but complete, farmstead. This place was later occupied by Rudoph and Lena Schendel. Korupp reported that he had enough work contracted to keep six people busy for the season.

C. A. Litchfield surprised everyone in June when he announced that he had sold the lumberyard plus associated lines and his residence to the Nitardy Brothers, of Iowa. The Nitardys immediately stated that they would enlarge the business to include a complete line of hardware. They planned to purchase two lots and erect a store building and a residence. The 1897 plat map of the village indicated that the second house was the one on the hill on the east side of the county road coming from the south. From other articles in the paper it appeared that Litchfield speculated in locating prospective communities and establishing a business with the intention of re-selling it. Within a month the Nitardys had a crew building the large lumber shed that stood in place until the 1980s. However, they indicated that they still had not located a lot for the hardware store. Then one of the brothers fell ill, and in January 1898 Otto Nitardy sold sub lot two in lot six to the Winona Lumber Company for $1,000. This lot was east of the stockyard, and the house was one of the more substantial homes in the village. The new company installed a telephone, which one could use to connect with neighboring towns.

Soon after the Nitardys had purchased the house on the south side of the village, Eva Fedder sold her lot with a small building directly to the east of the Nitardy lot to A. W. Peters. Peters apparently had rented the store on the west side of the county road from Wolter. He advertised that he was a dealer in dry goods, boots and shoes, groceries, hats, caps and more. The store was also a telephone station. Then, late in 1897, he announced that he was planning to build a new store along the street east of the Grandprey Hall/blacksmith shop. The building would be 24 x 50 foot two-story construction by Henry Korupp. The family would live on the second floor.

In January 1898, twenty-one teams were engaged in hauling stone from the Lindersmith quarry south of Clinton Falls. After the cornerstone was laid, refreshments were served and Peters stated that the building would be finished by early spring. On March 21-23, Peters and Waumett moved into the new store, which was built at a cost of $1,500. A news article read: "They have as neat a store as one can find in a small place. There is plenty of light [because of the high ceilings] and the interior is perfectly arranged." L. C. Wheelock drilled the well and built a seventy-two barrel cistern to complete the project. In July they had Prize Day. John Ruedy, of Woodville, won a set of decorated dishes; Louis Walthers a set of plain dishes; Henry Scholljegerdes a small scale; and Fred Olheft a phonograph. That must have been a busy spring for the partners, for in addition to moving into the new store they were also busy shipping out a carload of potatoes a week.

Within days after Peters opened his store, it was announced that in early April H. J. Luhman & Co. would open a new store in the building vacated by Peters. Luhman was from Preston, where he was employed as a clerk with Hard and Kuethe, who were backing him in the venture. Their ad read: "New Store New Firm New Goods etc. We propose to do cash business, but all kinds of farm produce will be accepted in place of cash at the highest market price….We hope…to merit a share of public patronage." Luhman's sister became a clerk in the store and had rooms above the stores, as did Luhman and Mr. & Mrs. Henry Rosenau.

Henry Luhman, general merchant 1897-1935, postmaster 1917-1928, c.a. 1910. Virena Palas photo.

Sam Grandprey visited experienced blacksmiths, especially one in Waseca, to learn some of the more difficult tasks and managed to grow the business with regular helpers until 1898, when he had the good fortune of securing the aid of John H. Franz. When he first came to Meriden, Franz was selling Maytag hand-powered washing machines, but he was not doing well because he was ill. Hattie nursed him back to health and earned a good discount on a new washing machine. She no longer had to use the scrub board. Hattie was happy and so was Sam. Business did exceptionally well from that time on, and because Meriden always seemed to be short of living quarters, Franz lived with them until 1905.

No doubt some of the good luck that Grandprey enjoyed was because on February 21, 1899, the Hocking blacksmith shop and adjoining dwelling was destroyed by fire. Hocking was pleased with the prompt insurance payment of $950. The Modern Woodmen of America, who had leased the upstairs hall of the building, lost all of their paraphernalia including some uniforms. The fire threatened the Peters and Waumett store, which stood directly west of the Hocking property. As soon as he received the insurance settlement Hocking left for Fairfax, where he built a new shop. He sold his lot to a Mr. Thone, from Faribault, who had plans to open a saloon. After Hocking left Meriden, Grandprey acquired his McCormick machinery dealership and sold plows and harrows, while Franz made hay racks, wagons ,and carriages.

In November 1899, the Winona Lumber Co. imported Kasota stone for the foundation of the new 32 x 32 x 16 foot building being constructed by Henry Korupp, who was also installing a full line of shelf and heavy hardware. My recollection of the contents of the hardware from the 1930s was that it included kitchen needs, flashlights, jack knives, builders' needs, and nuts, bolts, and harness for farm needs.

There was constant reference to E. L. Scoville and Anton Schuldt buying livestock and shipping to various markets. They had access to flat storage for grain and the stockyard the CNWR built as soon as it established the station. These men were reputed to be the major buyers within a ten-mile radius of the village. Shipments of grain and livestock went on throughout the year. The shrinkage on the animals must have been excessive. Think of the loss in this shipment. In March 1896, hogs were driven to the stockyard then loaded on the cars on the 14th, and on the 16th they arrived in Chicago. During one of Scoville and Schuldt's bigger weeks in 1898, they shipped two cars of wheat to Chicago, two to West Superior, and one to Minneapolis. On one Monday in July 1899, they shipped three cars of hogs and two of cattle, probably to St. Paul or Milwaukee, while the

Schultz Brothers shipped a car of hogs to St. Paul. During periods when grain prices were down, farmers did not deliver. When there was a hog cholera outbreak in 1897, shipments were at a standstill.

During the fifteen years covered in this chapter, the township attained its greatest number of farms and its peak in population. It also experienced the loss of its first major business, the Meriden Milling Company, which never recovered from the damage caused by the 1883 tornado. In 1886, its eighty-four horse-power boiler was sold to the Owatonna Packing Company. The company remained the largest and longest standing delinquent taxpayer in the township pending final legal maneuvers. In 1887, Palas, Barncard, and Waumett, the remaining founders of the company, brought suit against Henry Drinken to compel him to pay his share of the liability, but Drinken declared that he was not a stockholder and had merely signed a note as surety as an accommodation of the complaining parties. There was no future reference to the lawsuit, nor was the company again listed as a delinquent on the tax rolls. The remnants of that pioneer company, a large two-story red building, remained on the lot until the 1980s when it was razed by Eddie Kath. [2]

Daily Activities

Religious Events

As stated previously, this history will not include the history of St. Paul's Lutheran Church, but some of the external events of the congregation will be recorded. In October 1890, the congregation built the schoolhouse that served as the confirmation school and for other youth activities. It was finished by January 1891, and Rev. Kuethe stated that it "was a very prosperous addition to the church." In 1894, Henry Korupp & Son were contracted to build a new parsonage. That building was completed in October just in time to accept Rev. Wm. Nolting, who had been called from Harmony. Each year after confirmation the pastor and class traveled to Owatonna for their group picture. In 1896, the congregation purchased a lot 16 x 20 rods from Henry Palas for $50 to enlarge the original property. That year the Sunday school had a picnic at the Fred Altenburg farm. The year of expansion was completed with the construction of a new barn for the teams of the parishioners.

The St. Paul's Lutheran Church and parsonage completed in 1876. Virena Palas photo.

(This is extraneous, but I remember this barn had eleven doors. Each opened to a parti-tioned stall with a separate stall for two teams. The partition for each unit was about eight feet high so the horses could not reach over. The barn held forty-four horses. One day during confirmation school in 1937 we shut the door on one of our classmates on the north end of the building just before the bell rang to return to class. We left the door open on the south end of the barn so he had to climb over ten of the high partitions to get out. The minister suspected what we had done, and when our classmate came in late the rest of us boys received a good sermon!)

In 1898, services were cancelled so the congregants could attend a missionary gathering at Maplewood Park in Waseca. During the years covered in this chapter the congregation prospered, and at the same time it permitted its pastors to serve in congregations that were being founded in the area.

In the previous chapter, it was stated that the first services for the German Methodist Evangelical Church in the village were held in 1878. In 1885, the Minnesota German Methodist Conference had sixty-seven churches, which appears to be its high point. By 1924, it was absorbed into its English speaking counterpart. Apparently the Meriden congregation was one of the last to make the switch. In 1889, Rev. Petrich, a Meriden resident, served in the village and also filled in at Owatonna. Petrich apparently owned the store that was adjacent to the church. The next notable reference to the church came in January 1895, when it was reported that a revival meeting, which was held on the previous ten days, "would be continued for another week because it had created much interest." The next issue of the paper recorded: "The revival meetings held at the Evangelical church came to an end with good results."

It was nearly a year before the next reference of significance appeared. It stated that Sunday evening services were interesting and largely attended. "A collection of over nine dollars was raised to defray expenses of Mr. Cartwright who has been so ably filling in the pulpit of late." In June 1896, Rev. Sidow, from Mankato, arrived to be the new pastor. The following week many members of the congregation attended a camp meeting at Iosco (a township north of Waseca). There was little of note until 1898, at which time Rev. and Mrs. Englehart, M/M Henry Palas, John and Henry Walter, and Miss Matilda Peters attended the quarterly meeting in Waseca. During the week the Young People's Alliance elected officers of whom Will Schuelke was president; Miss Emma Peters, secretary; and Miss Ann Peters, organist.

In May 1896, the annual state conference of the German Methodist Evangelical Church was held in Meriden. The first order of business covered two days to examine candidates for the ministry. The next day was devoted to casualty insurance issues, followed by an afternoon on mission work. The conference concluded with the business meeting. "The conference was one of the best attended and most successful in church history. The...church is a strong organization and maintains several schools and a publishing house in the U.S. besides being active in world missions. The delegates were well cared for at various farms in the neighborhood." The local members must have felt good about having the conference meeting, for soon after it was over they purchased a new organ. But only two weeks later it was announced that Rev. C.C. Englehart and family were moving to Chaska, the Meriden congregation was assigned to Waseca, and Rev. Sidow would be the pastor. It was not until June 1899 that it was reported that the Young Peoples Alliance had an excellent program at the home of A.W. Peters. "There was a good attendance."

At the same time other groups were holding church services in the schoolhouse. In September 1895, the *Journal* reported that Rev. E. M. Calvin "who has filled the pulpit of the Presbyterian churches in Pratt, Bixby, and Meriden in the past few months" would return to his studies at Princeton. On nearly every Sunday during 1896, services were held for Baptists, Methodist, Presbyterians, and Congregationalists and were generally conducted by Owatonna preachers or by students from Pillsbury Academy.

Educational Opportunities

By 1890, the village had outgrown the first school building, and in October the second building was erected. This building, although modified and enlarged, was still there in 2012. It was a 24 x 38 x 2 foot structure and was furnished with Thomas Kane desks, good blackboards, and globes. It served the community for the next three decades. The major decision made in the 1890s was to adopt a nine-month term while most rural districts had eight-mouth or even seven-month terms. By 1899, forty-five were enrolled in District 48 under the guidance of Pearl Heath, of Minneapolis. She was greatly relieved when the parochial school at St. Paul's Church started, for many students went there four days per week and then to regular school on Friday. But she said she "had satisfactory attendance during the first four days."

The second schoolhouse with forty-four students and teacher. It faced north and was located at the southeast corner of Mill Street and the country road, where a garage was later built.

Having easy access to the railroad helped those who wanted to continue their education. In 1897, W. P. Canfield of Stillwater (Minnesota) Business College came to Owatonna to establish a business school. The original school was opened in October in the old Opera building. In 1883, Hattie Grandprey, nee Hersey, was probably one of the first in the township to attend Pillsbury Academy. Others may have attended, but in 1897 Lilly and Hulda Peters and Willard Buscho used the train to travel home most weekends. But those like James Andrews and Henry Rosenau, who wanted more education, found a way. In 1899, Wm. Brase Jr. became the second to enroll in the College of Agriculture in St. Paul to take engineering.

Social Life and Sports

Newspapers made every effort to maintain good local correspondents because that made it easier to sell advertising and subscriptions. News of the official proceedings from the courthouse or sheriff's office was important, but most people wanted to know what was taking place in their neighborhood and the larger community. People were not oversaturated with news like they are today. Some correspondents reported every event, while others were very selective about what they wrote.

Social life centered on visiting neighbors with card playing or dancing, always ending with a lunch at midnight or later. One of the most elaborate gatherings was probably that held by Mrs. L. W. Schnee, wife of the depot agent, when she had about seventy over for a lawn party. The grounds were decorated with Japanese lanterns, and a musical and literary program was held. Mrs. Schnee did oil painting and during her time in Meriden had many sales. When winter arrived she had as many as ten couples over for Progressive Cinch. In one of those Mrs. John Waumett took first for the women and Sam Grandprey first for the men.

As stated previously in this chapter, the CNWR did much to lessen the isolation. For example, many attended the Ringling Bros. circus in Owatonna in July 1896. Henry Walthers attended the national encampment of the Grand Army of the Republic in St. Paul. That same fall nearly every train had someone going to the state fair. A special tourist train provided round trip tickets for $0.45 to a festival at Maplewood Park in Waseca. That November many of that group took a sleigh ride to Waseca and returned by the moonlight.

Being late for the train proved costly for a party of eight that attended a play, *Dodge's Trip to New York*, which was performed in Owatonna one March night. By remaining until the end of the play they all missed the train. Four of the group secured a livery team and drove home, and the other four remained in Owatonna for the night and took the train home the next day. In December 1898, as they had done previously, Mrs. C. H. Wilker, Mrs. John Scholljegerdes, and R. H. Chapin left on the CNWR for San Diego for the winter. Members of the Meriden Orchestra—John Franz, A. W. Peters, Ferdinand Peters, Merton Parcher, Henry Korupp, and Charles Markmann—had music instructors come out on a weekly basis to receive music lessons.

In the village this pattern changed after 1896 when Grandprey Hall was constructed. The hall was on the second floor over the blacksmith shop and quickly became the social center of Meriden. The township also rented for official functions. As stated above, the Hocking blacksmith shop also had a second floor for public events. During the winter months there were often two events each week. For example, in one January event a combined raffle and dance at Grandprey Hall had music furnished by Sam Grandprey and John Franz. Herman Ahlers won a bicycle, John Mathews a rifle, and Carl Rayfeldt a revolver. Two nights later a dance and party were held in honor of Mr. and Mrs. J. E. Bryant, of Minneapolis, and a large number of both young and old attended. On the same evening the Woodmen lodge held a meeting in their hall and installed new officers, after which the ladies held a surprise supper followed by dancing. More new members were initiated into the lodge the following week at which time an insurance check was paid to Mrs. A. E. Powell on the death of a relative. At another meeting Albert Rosenau sponsored a dance and raffle at Grandprey Hall. The Woodmen provided the music for the thirty couples attending who also partook of a supper at the Meriden Hotel (aka Parchers Hotel). Admission was usually $0.50 a couple, but often a person or group sponsored the event. Many wedding dances were held in Grandprey Hall. Sometimes small groups would rent the hall for dancing, and to save money they used a gramophone for music. Traveling events such as a Punch and Judy Show with performing dolls made up the program, and at others Chief Broken Branch and Seneca Indians performed tricks while a white man sold medicine. One time an Easter dance was held at the Hocking hall, but it was advertised that "no liquid refreshments were allowed in the building."

The most serious business at the town board meeting in May 1886 was that county commissioner L. L. Bennett (president of the Farmers National Bank) reported that the license to sell intoxicating liquors in the village would be $100. There were references to a saloon in the village at various times, but in most cases the only liquor sold over the counter was beer.

The local correspondent in 1889 lamented, "Years ago we supported two hotels and both died out about the same time and we need one for travelers." Two years later the complaint was made again: "This place is greatly in need of a hotel and boarding house." Joseph Grandprey apparently revived the hotel, and in October 1896 rented it to Charles Depoe of Owatonna, who ran the business until January 1898 when Sarah (Mrs. Merton) Parcher, Grandprey's daughter, operated it under the name Central Hotel. The following March he deeded the property to her. Soon after a 12 x 18 foot addition was built on to serve as a dining room.

There were horses and drays in the village, but it was not until 1895 that any direct reference was made to a livery stable. The barn for the livery stable was located south of the tracks next to the combination store/hotel building. The only account other than hearsay from older folk I encountered was in the May 24 *Journal*, which stated: "Dan Shoemaker, of Owatonna, took a horse and buggy at this place for the remainder of his way."

Probably the innovation that was close to being a craze during this period was the bicycle. Bicycles are a British invention that came to America in 1878, and by 1900 over 70,000 workers were employed in making a million bicycles that were sold during the decade. Meriden was not left out, for in March 1896, L. G. Wolter and John Palas traveled to the Twin Cities to secure a bicycle dealership. The first bicycle ad appeared in the *Journal* that week. By April, Walter had sold eight from his store, including one to H. J. Rosenau and two to the children of A.W. Peters, the children of his competing grocer. By May several had ridden on their bicycles to Owatonna and returned. Mrs. Mattes rode hers to Owatonna, but she said she "was very much fatigued. . . so she visited over night with Mrs. Henry Mundt and rode back the next morning." In June, Wolter and Peters bicycled to Owatonna, Faribault, Morristown, Waseca, and Meriden, about seventy-five miles. "They found the trip enjoyable and roads in fairly good condition." In February 1898, twelve from the village bicycled to Owatonna to attend the big masquerade given by the Elk bicycle association.

Twelve prominent individuals, many from outside the village, were charter members of the Meriden Gun Club who purchased a trap so they could have regular shoots. My father had a trap in the 1930s and said that it had been passed down over the years but did not know if that was the original. The group had several shoots, including contests with other clubs, each year, even during the winter months. Albert Rosenau, who was single and appeared to be a man-about-town, went on a hunting trip to Le Sueur County in October 1894 and returned with fifty-seven ducks, thirty-six muskrats, and twenty-five squirrels. Each winter several of the locals went

to the northern part of the state to hunt deer. In November 1899, "they got
no deer but came back in good health."

Without a doubt fishing rated the number one sport along with baseball played in a convenient local pasture. There was always a ready supply of bases. At times Meriden had a formal group that played Blooming
Grove, Deerfield, Lemond, Woodville, and the other townships. If a team
was good, they always complained they wanted to play games on more days
than just Sundays.

Communication, Crime, Health, Et al

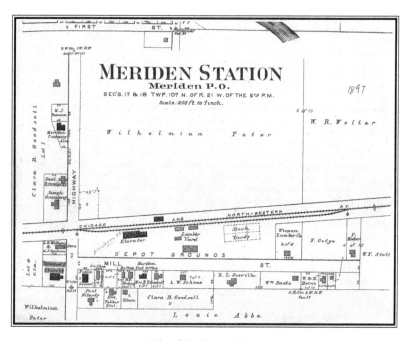

Plat of Meriden in 1897.

Most of the people of the township never experienced any prolonged
isolation, for by 1866 the railroad and the telegraph connected them to the
outside world. Owatonna had a limited telephone system by 1883, and in
1896 a rural line was built through the township and into Waseca County,
but it was not until 1897 that the Peters & Waumett store and the Winona
Lumber Company had telephones. Both firms were connected with the
Owatonna system. Basically, it meant that except for difficulty in hearing
because of the poor rural connection, long distance phone service was available. However, prior to the late 1930s long distance calls were rarely made.

Rural store owners liked being postmaster. Even if the pay was not enough to live on, it was a boon to the store because it helped to draw customers. But in 1892, Postmaster General John Wanamaker pushed the expansion of regular mail delivery to the rural areas. The government had conducted a seven-month experiment in forty-six rural offices and showed a profit of $850. Postage at that time was one cent for a first-class letter. During 1898, Waseca put rural delivery into service, and by late 1899 rural patrons in the Meriden area were informed of the possibility of it coming to the area.

Public meetings originally were held in the depot and later in the post office, but in early 1898 it was voted that the polling place be moved to one of the halls. The vote was thirty-four for Grandprey Hall to twenty-six for Hocking.

As the village became more established, the citizens became discontented with the dirt (not graveled) streets and the lack of sidewalks. On October 14, 1898, a meeting was held that resulted in a petition to get sidewalks on the main streets. To meet part of the cost a dance was called for October 28; a supper at the hotel was also held. By November a path was graded for the sidewalk to run on the south side of Main Street beginning with the schoolhouse east to Mrs. Goodsell's store and post office (a distance of about 100 feet past the Peters store).

It was proposed that it should extend north to the depot. That was never cemented, but a board walk was built. A second sidewalk was projected on the west side of the county road from Luhman's store to the hotel. Meriden had its first sidewalk, and after November arrived the big concern was who would shovel the snow from the sidewalk.

The township was not without people who stepped outside of the law. For example, on June 22, 1887, Arthur Sheldon of Waseca was charged for stealing a horse from Frederick Walter of Meriden on June 13. His bond was set at $500 and he was scheduled to appear at the next district court. Bachelor Albert Rosenau, whose hunting and trapping ventures were mentioned previously, attempted to attract attention another way. In July 1893, he went to a circus in Owatonna dressed as a policeman so he could get in free. After the main show, on the way out he arrested a drunk and threatened to take him to jail. The drunk resisted, but after paying Rosenau $2.50 he was let free. The Owatonna marshall learned about it and arrested Rosenau and fined him $10. Meriden justice F. W. Goodsell heard the case of F. Enders vs. Mr. Jourgerson in which Enders complained that Jourgerson had killed his chickens. Goodsell ruled Jourgerson was guilty and rendered a verdict of $1.00 and costs.

The depot was a favorite target of vandals. In May 1892, after securing tools from the blacksmith shop, two "brethren" forced the door of the depot and drilled three holes in the safe. The handle and the dial were broken off. "Just why the attempt failed cannot be told with certainty. The drill was broken off in one hole and struck a chilled steel plate in another. One of the burglars exchanged coats with agent Bryden. The safe was shipped to Chicago." In November 1894, burglars broke into the CNWR tool shed and took tools and the section car. The next morning the section car was found on the sidetrack opposite the State School in Owatonna. In July 1895, someone entered the L. Walter store and took cigars and dry goods. The money drawer was also removed, but Walter had taken the contents home when he closed for the day.

James W. Andrews, whose family arrived in 1856, he always been active in both Meriden and Owatonna. He was born July 22, 1867, and after his education at Pillsbury Academy he taught school, including one year in the village. In 1900, he ran for county auditor and won by sixty-two votes. He held that position from January 1901 through December 1904, after which he worked in the Owatonna post office until 1917, when he joined the family in farming.

James W. and Katie Andrews. James W. Andrews photo.

The last year of the century was a banner year for the village. For Contractor Henry Korupp had a crew of ten engaged for the entire construction season. At least five houses were built in addition to a set of buildings on the Fred Fette farm. These were substantial houses and built by people who were moving from the farm and were renting their land to others or turning it over to the next generation. One house was for Wm. Patten who, according to local lore, was a Civil War veteran and was suspected of being a deserter since he never applied for a pension.

When people were ill it was satisfying to know that Owatonna and Waseca doctors made house calls either by taking the train to the village or, frequently, they had their own team and buggy and traveled directly to the village or farmstead. Probably the saddest event in this regard was the death of John Buelow, age forty-eight, who died of apoplexy, leaving a wife and seven children. Other deaths were caused by consumption, indigestion, diphtheria, measles, and typhoid fever, but the township continued to grow because births far exceeded deaths. [3]

---

[1]   Robert J. Casey and W. A. Douglas, *Pioneer Railroad: The Story of the Chicago and North Western System* (New York: McGraw-Hill Book Co., 1948) 217, hereafter Casey & Douglas; Drache, *Challenge* 233–234.

[2]   *Journal; Chronological Landmarks in American Agriculture: USDA Agriculture Information Bulletin No. 425 (1975), hereafter Chronological Landmarks*; Drache, *Challenge* 5; Hattie Grandprey, A *Memoir*; Lloyd Grandprey Jr., telephone interviews, various dates, hereafter Grandprey Jr interview.; Lloyd Grandprey Sr., personal interview, 11 September 1976; Paul A. Drache, personal interview, 11 July 1976, hereafter Drache interview; Robert Schultz, telephone interviews, 16, 20 May 2001, hereafter Schultz interview.

[3]   *Journal*; Harold Williamson, "Mass Production for Mass Consumption," *Technology in Western Civilization*, 686; James W. Andrews, letters, 25 May 2005, 14 November 2008, and several phone conversations, 2006–2010, hereafter Andrews.

CHAPTER IV
Still Developing 1900–1914

The Chicago & Northwestern

Although the WSPR and the CNWR had been working together for several years it was not until 1900 that WSPR was officially taken over by the CNWR. This enlarged system led to continued upgrading of the facilities and expansion of service. The established schedule, in addition to the local freight, was three passenger trains east and three to the west each day.

Beyond the service that it provided, the CNWR was always an important employer in the village. In addition to an agent it employed a section foreman and nearly always three regular hands; from April until freeze up it was not uncommon to employ an additional four workers. The CNWR pay scale was above that of most of the other people working in the village, which proved to be a boon to the local economy. In 1909, the Western Union had a crew of line men setting new poles to upgrade the telegraph service, which meant the local hotel had to provide twenty-one dinners daily.

Contrast Meriden's advantages to that of an inland community like Lemond where the creamery had to haul its butter eight miles cross-country or at best on dirt roads with teams to deliver its weekly production to the freight car. Or when a twenty ton or greater car of coal arrived in Meriden and twelve to fifteen farmers, with teams, had to be assembled to unload the car in time to avoid demurrage. Such handicaps made it almost impossible for the creameries and merchants in the inland communities to compete.

Meriden residents appreciated the service and made good use of the railroad. For example, on one day in March 1901, nineteen people traveled either to Waseca or Owatonna and back again. On July 4, 1904, eighty-five people purchased round-trip tickets to Owatonna where they went to celebrate the day. On July 4, 1907, the train was flagged down by the agent to accommodate forty people who went to Waseca to celebrate the events. In September 1907, a party of fifty went to Owatonna to take part in the street fair. By 1910, an average of twelve students made round trips each day to attend OHS, Pillsbury Academy, or Canfield Business School. In addition, five or more business people or those going to town for medical or shopping needs made the trip. In January 1912, four shipments of butter totaling 111,664 pounds, about thirty cases of eggs from each store weekly, six cars of grain, and twenty cars of livestock were exported and two to four cars of freight were received weekly. On July 4, 1912, sixty passengers boarded the early train for the big celebration in Owatonna. The above examples were considerably above average, and it took the fare from several passengers to just pay for the cost of stopping and starting the trains, which meant that the passenger business was not profitable. The freight business had to provide most of the income.

A curious side light to the village situation was that it always seemed to have a housing shortage, especially for families. By this time the proximity to Owatonna and Waseca would prevent it from becoming any larger. Post offices in Havana, Merton, Omro, River Point, and Steele Center had already been discontinued in the county.

However, as the population in the larger communities increased and new stations were founded further west along the line, the railroad continued to grow. In 1903, to keep up with the demand, the road started to lay larger and heavier rails. The grade was lowered in Meriden to make it easier to load and unload the cars. In 1904, the siding was ballasted with fifty-eight carloads of cinder. At the same time, the depot was improved and enlarged. By 1906 heavier rails were laid on the entire main line, which meant that larger and faster trains with heavier loads could be put into operation. The engines required more coal than a fireman could shovel, so the company announced that it would install 700 mechanical stokers that year and complete the installation process in all the engines in the following years. The next year all of the bridges from Winona to Waseca were strengthened.

Anything that happened along the road was news. In July 1903, Henry Drinken and Henry Korupp were on the train to Owatonna that was involved in a wreck. Apparently, neither of them received any injuries. When a freight train went through the village at 1:00 a.m. on September 23, 1904, the engineer spotted that Frank Eggert's barn in the village was on fire, and he blew the whistle, waking everyone. Soon a crowd was on the scene, but they were too late to save the barn, a horse, a buggy, a sleigh, two sets of harness, and the feed. The *Journal* article closed, "But their valiant efforts saved the house," which remained a home for many years. In October sparks from a passing train engine started a fire that destroyed the Pride elevator in Havana. Because the elevator was so close to Owatonna it was not profitable to maintain, so it was not rebuilt. In May 1907, sparks burned a stack of hay just east of the village belonging to Gust Fette, "for which the railroad settled immediately." In November 1907, about a mile and one-half east of the village, the trucks broke out from under a car of coal on a west bound freight. The train was split in two with the front end going to

Meriden and the rear of the train pulled to Owatonna. Traffic was delayed for more than ten hours before the car and contents were removed. In July 1910, a ditch fire caused by train sparks threatened to burn Luhman's barn and "could have threatened the whole town but village residents worked furiously and stopped it." In November 1914, the wooden trestle two miles east of the village between sections fourteen and fifteen was burned because of ditch burning. Local passenger trains from both sides met and exchanged passengers and baggage, after which each train backed to Owatonna and Waseca respectively. All the freights and passenger train #514, the through train from Chicago, had to detour through Faribault-Waterville-Waseca and then west or vice versa when going east. The final catastrophe for this period came when five horses owned by the Waumetts, who lived more than a mile from the tracks, were killed by a passenger train.

The Business District

Post Office and Telephones

The government had experimented with rural free delivery (RFD), and by 1896 it proved to be successful, so the government decided to expand the system. In November, the *Journal* carried a notice that gave the details. Under ordinary conditions it considered that a route should be twenty-five miles. The route would be devised in such a way that a carrier would not have to pass over the same road twice a day. The pay was set at $400-$500 per year, and other than good character there were no civil service or gender restrictions. In 1901, four rural routes were established out of Owatonna. The first route was called the Morely (an early settler) route through Meriden, and it was to be served by Sidney Glasgow with W. E. Glasgow as substitute. The annual salary was $500. The postal department notified rural patrons that they had to display a sign if there was mail to be sent because the carrier was not required to stop if there was no mail to deliver.

It is a well-known fact that small community stores disliked RFD because it meant they would have to compete with firms in larger communities, especially mail order firms. Many country stores survived because they had the post office. Sam Grandprey broke that precedence in Meriden on March 26, 1904, when he was named postmaster and the post office was removed from the A. W. Peters store. Grandprey purchased the Goodsell shack from the lot north of the creamery and moved it to the west side of his blacksmith shop. John Franz was named assistant postmaster. The building was remodeled to accommodate a telephone booth and a place to serve as library. The Owatonna library had made arrangements to provide books to the village with Rachel Peters in charge. Grandprey also planned to carry a line of cigars, a confectionary, and patent medicine.

A 1900 photo of the Samuel E. Grandprey family.
L-R: Samuel, Lloyd, Medora, Harry, Helen, Hattie.

While the above was taking place, the Tri-State Telephone Co. unloaded eleven carloads of telephone poles and two carloads of wire, which was enough for sixty-eight miles of poles and wire to coincide with the construction of exchanges in both Owatonna and Waseca. By July 1906, the rural lines were being used to keep farmers posted on important news. A special long ring would be sent at noon and again at 7 p.m. at which time the market news was given along with general news including news of the world. Some farmers placed their grocery orders by phone before coming to town. William Walthers, a Meriden livestock buyer, ran an ad stating that he had installed a Waseca phone "so farmers can call him any time." In the July 19, 1907, *Journal* the editor commented, "More or less opposition was at first met in the county toward the rural telephones, but gradually the rural residents have become convinced that the telephone is a necessity and a great time saver…now they are anxious to obtain the service." By then the North West Telephone exchange out of Owatonna had forty-five subscribers from the village and immediate area. The system had grown so much that the Meriden Rural Telephone Co. (MRTC) was founded January 5, 1913. A. T. Schuldt was elected president and H. H. Wicklow secretary. The board funded the cost of erecting the telephone line and worked on extending the line "as rapidly as possible." Within a month they had thirteen patrons. In March another local line was merged with MRTC. At a meeting in October it was determined that a subscriber, i.e. stockholder, had to pay $65 for two poles and a phone. From then on the subscriber would be assessed $1 annually plus any long distance calls, which were a rarity at that time. Renters of land could rent a phone for $12 per year; the company would provide the phone and one pole. If more than one was needed, the renter had to pay for that in advance. John Burr was assigned to take care of the east end of the line with the 730 prefix numbers and A. Matthew the west end with the 723 numbers prefix, which included the village. In April 1914, the organization held its end-of-the-year meeting at the court house. Within weeks after the village and its trade area phone lines were completed, lines were erected in the Crane Creek area.

During those years most farmers were getting a weekly paper in addition to farm magazines and probably a church magazine. In the case of my maternal grandparents the church paper, *Das Kirchenblat*, was in German.

By then there were ten rural mail routes out of Owatonna. Routes nine and ten passed through the township, one from the north and the other from the south. They circled back on the roads north and south of section seventeen to take in the village. In 1911, all of the routes were increased to over twenty-four miles, and the carriers' pay was increased to $1,000. The volume of mail had increased even though parcels were still not carried by RFD.

John Ebeling, the Meriden delegate on the County Good Roads Association, reported that the county had 270 miles of postal routes that were traveled every day. He warned that mail service would be cut off if the roads were not improved. The Association had discussions about taxing for roads rather than relying on the work being done by the taxpayers via the poll tax. The problem continued, and in 1913 rural carriers protested the conditions of the road and identified specific spots on every route. The government decreed that service would be stopped over any route that was not being kept up. The problem was compounded because relying on local residents to do work in lieu of taxes was no longer satisfactory. "Carriers were upset by the attitude anything is good enough for the mail man."

General Merchandise Stores

A 1903 booster edition of the *Journal* read: "Meriden lies in the finest farming district of the county and the village itself is one of the prettiest little towns in the state. It has a population of about 200, the great majority of the residents being of sturdy German parentage. Its people are wide awake and progressive, and every day of the week is a 'busy day' for Meriden." The only negative was that it had no hotel, "but one was badly needed." The article continued that the creamery was the star creamery of the state because H. J. Rosenau had a score of 98 in July. The east side of the creamery had a large banner sponsored by the state dairy association proclaiming "highest score." In the days before electricity, if the grocer wanted to sell soda pop, beer, butter, meat products, or ice cream, it was necessary for him to have a large ice box, which meant that he needed an ice house. Both grocers in the village had ice boxes, and each year they had to haul ice from Goose Lake, if its water level was high enough to provide the ice. But most of time, like the creamery, they had to travel the extra three miles to Clear Lake. In either case, it was a time consuming and labor intensive task, but having products from the ice box was a "real treat," since most of their customers did not have a cooling appliance.

The grocer also sold kerosene, which at that time was the major product of the petroleum industry. It was relatively inexpensive and an easy-to-use source of light, except that the housewife had the daily task of cleaning the lamp chimneys. The grocers sold kerosene because, unless the blacksmith or the lumber yard/hardware handled it, there was no one else. In August 1903, the horse-drawn Standard Oil Co. wagon from Owatonna came to the village to make deliveries to the stores. This is the first reference to the company that later attempted to make Meriden a bulk station. In 1904, Harry Stout, who I had the pleasure of knowing in his later years,

A typical 400 gallon Standard Oil horse-drawn delivery wagon, 1913.

called on the merchants to present his full line of cigars that the stores and the pool hall carried. It was reputed that the post office also sold cigars.

In February 1904, A. W. Peters surprised everyone in the community when he traded his store and inventory for land in Clark, South Dakota, but he had not decided what he would be doing. In April, the Lund Land Agency started a closing out sale of the store. There was to be a public auction every Wednesday and Friday afternoon and evening to sell the complete line of goods of the store. The building would also be sold. To appeal to the short-of-cash customers, they offered the highest price for eggs.

Henry H. Wicklow, general merchant, 1907-1945. Dorothy Wicklow Hudson photo.

In June, Mr. & Mrs. C. B. Hanson moved into the living quarters of the store to continue the liquidation. In November, Henry Wicklow purchased the building for $2,000, and shortly after he and his mother moved to the living quarters. Wicklow remained as clerk in the Luhman store until April 1906. It is of interest that his daughter Dorothy had a letter from Henry Luhman, dated April 7, 1906, which stated that Henry Wicklow had been in his employ for five years and he "was pleased to testify of his ability and good character. He leaves our company voluntarily with our best wishes." On April 27, 1907, Wicklow opened his general merchandise store, which became one of the village's most successful and long-tenured businesses.

The H. J. Luhman store purchased a new dray to haul eggs to the depot and to secure groceries that were shipped in. Luhman also purchased a new buggy for personal use but also to make deliveries. In March 1905, his store shipped sixty-five cases of eggs (thirty dozen cases) to Chicago on a regular basis, which indicates that many farmers were keeping sizeable flocks.

Wicklow also handled eggs and chickens, and in 1911 he also shipped live chickens to the Chicago market. He offered $0.101/2 for spring chickens and $0.08 for old hens and for roosters. Like the grocers before him, Wicklow sold potatoes to local customers, but he also brokered carloads of potatoes for the Chicago markets, selling as many as two carloads some weeks.

15 Mill Street 1907. L-R: H. H. Wicklow store, blacksmith shop with Grandprey Hall on second floor, post office, which was the original house on the Goodsell farm. Note the telephone sign on the post office and the cement sidewalk.

In 1913-14, both Meriden stores made changes to keep up with changing times. Luhman built a new barn south of the tracks to keep a team for his dray wagon and buggy. Then, he erected a galvanized storage structure that was adjacent to the railroad and county road. This was used primarily for salt and flour storage. Both stores handled block and sacked salt for animals as well as salt, sugar, and flour for human consumption, which came in 100-pound sacks. My parents purchased sugar and flour in 100-pound sacks as late as 1946. Wicklow had his salt and flour storage in a lean-to built on the west side of his store.

In 1914, Wicklow installed a "new Bowser Gasoline outfit . . . which is very convenient for automobilists." Most places that sold gasoline kept it in above ground barrels from which it was pumped into cans. Funnels were used to fill the gas tank. The Bowser was a hand-operated pump that transferred the gasoline from a storage tank.

In 1911, the Miller Bros. of Lemond purchased lumber from Hayes Lucas to build a 24 x 32 foot store in the village. It is hard to visualize that such a venture could succeed but it was still open in 1940.

A 1912 photo looking west. L-R: Wicklow store, blacksmith shop, post office (school not visible), the German Immanuel Methodist Church, horse barn, Luhman store, barn, L. G. Campbell Milling Co., Sam Grandprey manager, and in the foreground the boardwalk from Mill Street to the depot.
Mary Stendel photo.

The Blacksmith Shop and Hayes-Lucas Lumber Co.

When Sam Grandprey was named postmaster he realized that he could not continue with his blacksmith business together with his other interests, so on June 20, 1904, he made Charles Markmann, an employee for two and one-half years, a partner. That partnership lasted until 1917 when the Grandpreys moved to Owatonna, but the business continued to remain a key part of Meriden business district.

In 1900, the Hayes-Lucas Lumber Yard was under the management of Herman Martin and was still undergoing expansion. Area farmers were enjoying the golden age of agriculture and continued to upgrade their farmsteads. It was a good period for Hayes-Lucas. The yard often received two or more cars of lumber during the week, a similar amount of tile for tiling, which was a key to good farming, plus shipments of posts, cement, fencing, and coal. In 1904, Martin supervised the unloading of seven cars of cinders, which greatly improved the condition of the yard. In 1906, a lumber shed, a shed for cement, and a coal shed were erected. The coal and cement sheds were adjacent to the siding so their contents could be unloaded directly into them from the train cars. In 1909 a 34 x 76 foot lumber shed was erected directly south of the coal shed. This building was one of the favorite places for the youth of Meriden to play hide-and-seek. In 1914, the yard reached its full size when a 60 x 30 foot shed was built to handle farm machinery, and a 20-foot addition to the hardware department was erected.

Creameries

In June 1900, the Meriden Creamery Association started operation in its new brick and cement modernized building, and within a few months the board decided that it needed a new, larger ice house to replace the one they had moved in from the skimming station. The new structure, which had a capacity of 900 cakes of ice, was filled for the first time in 1901. In 1914, an addition to the plant included an ice plant, which made it easier to maintain a consistent quality product.

The new creamery had the very latest equipment and a good butter maker. Rosenau scored 96 in the 1901 national butter contest but received third place because the butter had a "slight deficiency in flavor." In 1903, ,the Meriden creamery purchased a pasteurizer for its cream and/or its milk and in October sold its first pasteurized milk. Soon after it shipped its first pasteurized butter to New York. The creamery continued to sell to New York and Philadelphia markets for many years.

Nearly all of the early creameries were small, wooden structures where it was not easy to maintain sanitary conditions. Probably the same could be said about the facilities on the farms from which the milk or cream came. In 1903 Steele County had twenty-one creameries and one skimming station. Only five of those creameries were built of brick or had sawed stone floors. They were very modest structures. For example, in April 1905 the Crown Creamery burned and by June that year it was back in operation.

In 1903, the 1,980 Steele County farmer patrons milked 15,500 cows, which produced an average of about 4,300 pounds of milk per year with an average of 3.78 percent butter fat. Only 107 farmers had cream separators, but somehow all involved were doing a good enough job that the county butter carried a premium of one-half to one and one-half cents per pound. In 1911, W. Wobbrock of Golden Rule scored 96.33, Rosenau 94, and five others from the county scored in the excellent range. Only three years later, the twenty-four Steele County creameries paid out $955,174 for butter fat. Only Freeborn, Ottertail, and Stearns counties had more creameries.

The Elevator

In the early years the grain business was never as successful as the creamery business. Most of the time E. L. Scoville was the only grain buyer, and for many years was the major buyer of both grain and livestock. In the first two weeks of December 1901, he shipped five cars of wheat, two cars of hogs, and two cars of cattle to Chicago. But business in general was slow. In April 1903, Scoville owned the old flat storage house, and the new tower elevator was owned by the Hastings Milling Co. Both were closed for lack of patronage, as was the hotel. Only the new creamery seemed to be prospering.

In August, in anticipation of harvest, it was rumored that the Waseca Millers would try to open at least one of the structures, but nothing happened. In September they tried again but were not successful. Then the Hastings Milling Co. opened under the management of John Franz, who had worked with Grandprey in the blacksmith shop. At that time it was disclosed that the reason the elevators were shut down was that the company was going through financial restructuring. The Meriden elevator was reopened as the Pride elevator, which apparently was a subsidiary of Hastings Milling. Most of the wheat from Meriden was shipped to them at Owatonna.

In May 1905, Franz reported that some days' receipts at the elevator were as high as 1,200 bushels. If the farmers had used double wagon boxes to haul their wheat, that would have required twenty loads. Today, that much wheat could easily be delivered in two truck loads. Business continued, and in August a 20 x 24 foot feed store was built next to the elevator so it could deal in livestock and poultry feeds. Franz resigned to go to North Dakota to search for land. Even though he held the position of postmaster, Sam Grandprey was appointed the new manager for Hastings. In his upbeat manner Grandprey reported that business had been good in his first week, and on one day he had received 600 bushels of wheat for which he paid $0.87 a bushel.

The Hastings Milling Co. apparently was a subsidiary of the Sheffield-King Milling Co. of Faribault. B. B. Sheffield sold his interest in that line of southern Minnesota elevators but retained his interest in a line of sixty-seven elevators and a terminal elevator in Minneapolis. Mr. King, the other partner, retained seventeen elevators, which included Meriden and mills in Morristown and Faribault. The Sheffield name was retained because Sheffield was reputed to be known worldwide in the flour trade. Meriden may have had some prospect of being a good grain trade center, for in 1907 the Van Dusen-Harrington Company, a large firm, announced that the flat storage facility would be disposed of. "It will either be sold or torn down." Within a month the company decided that "the elevator a landmark for years" would be dismantled and the timber would be shipped to Vogel (Volga), South Dakota, and rebuilt.

Sam Grandprey continued as manager. In September 1910, he announced that he had already shipped out his fifteenth carload of wheat that season. By that time it appears that the elevator was again the property of the L. G. Campbell Milling Co. Campbell was a hands-on manager and immediately repaired the foundation of the building. He lived in Owatonna, as did his workers, and each day he drove out with the crew in his automobile. Lloyd Grandprey, Jr., a grandson of Sam, stated that his grandfather suffered a back problem from working on that foundation repair project. The Campbell Company was a successful firm and later became known for Malt-O-Meal, which was produced in Northfield.

Probably the most unique business venture in the community appeared in 1903 when Ed Bonnell opened a feather renovating business. Initially, he rented space in Grandprey Hall and later moved it to the Reifeldt residence. His work consisted of renewing feather ticks and mattresses that were filled with feathers. This business was probably of a very itinerant nature, for Bonnell was only mentioned twice in the paper

Livestock Buyers

The CNWR drilled a well and put up a water tower so it had water for the engines. At the same time it built the stockyard so the farmers could deliver their livestock to the public markets. E. L. Scoville was probably the first buyer in the community for both grain and livestock and had the longest term of service in that work. In the first week of February 1901, Anton Schuldt and Wm. Walthers shipped two cars of hogs on Monday, and Scoville & Donovan shipped two cars of hogs on Thursday. The people at the yard had to contact the railroad when they needed a car or cars. If the road had no cars available, the yard would avoid buying livestock until they could get cars. In April, Schuldt and Walthers shipped three cars from Meriden and were forced to drive 100 cattle to the yard at Owatonna that was served by another road. In the last week of December 1901, three cars of livestock were shipped to So. St. Paul and three cars went to Chicago. In February 1904, John Scholljegerdes delivered fifty hogs four and one-half miles by driving them to the yard where they were loaded for the thirty-hour trip to Chicago. Walthers accompanied the shipment to Chicago.

Apparently Scholljegerdes had no trouble getting his hogs to Meriden, but that was not the case when E. S. Tuthill wanted to drive a carload of cattle to the stockyard. His helpers, Paul Gasner and Hans Halvorson, experienced a storm as they were driving the cattle into the village, so they put the cattle in Goodsell's barn located north of the creamery and immediately left for Tuthills. On the way the horses became confused and wandered off the road. The two spent the night wrapped in heavy blankets. At daylight they were able determine their location and drive home. Train service was halted for two days because of the storm, so the cattle did not get their regular feed during that period. This gives a good idea of the irregularity of shipments, how dependent everyone was on the railroad, and why farmers lost so much because of shrinkage. Those who accompanied the livestock to market rode free in the caboose, but the lost time and expenses had to be considered. In 1905, the Austin Packing Company was mentioned for the first time. But it was several years before farmers were comfortable with direct marketing, so there were no shipments to either the Austin or Albert Lea markets.

Meriden farmers wanted to improve their marketing system, so on February 15, 1910, they gathered at Grandprey Hall to discuss establishing a cooperative livestock marketing firm. Peter Moe, Fred Fette, and August Schuldt were appointed to make an investigation. The State of Minnesota gave approval for incorporation, and the Meriden Farmers Cooperative Company was organized on March 29, 1910. I have the stock certificate for three shares issued to my grandfather, Max Drache, in payment of $30. Everything proceeded rapidly, and by April 22, Fred Fette, who was acting manager, shipped the first car of hogs. At a meeting held in August it was stated that membership was continuing to grow. At its annual meeting in 1911 the organization had 182 members and everyone was assessed $3.44.

Business grew rapidly, and the common complaint was that they could not get cars fast enough. In February 1912, a new office was erected so the secretary/manager, Fred Fette, and the buyers would be able to spend more time there. This little yellow building was chiefly for writing out the transaction slips. However, it did not contain the scale beam so the manager had to weigh in the yard and do his computations in the office. After the bank was established the check writing was generally done there. In 1914, the Waseca telephone was installed.

Crane Creek–Handmaid to Agriculture

In a previous chapter Crane Creek was mentioned as being one of the few obstacles the township had relative to agriculture, but farmers soon took action to acquire an improved drainage system. In September 1906, Judge Blackham denied the application of the farmers' drainage petition. He said that the plans and theories advanced by the engineers could "only be demonstrated by actual trial . . . that would involve so large a cost he deems it inadvisable." The article concluded that a lawsuit might grow out of the issue. The expenses for surveying and preliminary costs by that date for engineers, viewers, attorneys, and court costs had exceeded the bonded limits, and the petitioners had to raise additional funds before they could proceed. Their efforts were rewarded, and in May 1907 the State awarded

$5,000 for the "famous Crane Creek Ditch." The state estimated that for each dollar spent on ditches it would receive two dollars in tax assessments, so Crane Creek was a high profile project. The state drainage commission gave its full approval. The local committee representing Steele and Waseca Counties consisted of James Andrews, J. M. Diment, Peter Moe, John Moonan, and Frank Domy. With that news, John Scholljegerdes, of Meriden, had a crew digging a ditch and tiling land in section five that would lead to draining Bradley Lake.

In the fall of 1907, the case reopened and portions of the northern and western sections were removed from the project. Soon after that Judge Blackham ordered construction of the Crane Creek drainage ditch to be officially known as Judicial Ditch No. 1. (See the plat map on the inside of the book covers.) Hayes Lake, Swan Lake, Pelican Lake, and Bradley Lake were considered meandered lakes with insufficient depth or volume to be of any use as fishing, boating, or a public water supply. There was no opposition.

In April 1908, the engineers reported the results of the final survey, and the estimated cost was $100,000 on what was claimed to be one of the largest ditch projects in southern Minnesota. There were still objectors to the project, and the State Supreme Court granted them a writ of certiorari, which meant that the District Court had to submit all of its proceedings to the Supreme Court for review. It was the only recourse that the objectors had. Bids were opened, but the objectors made one last attempt to block the start of the digging. They failed, and by August 1909 the dredge was moving closer to Bradley Lake in sections five and six and part of two sections in Deerfield. The other water bodies that were involved were in Blooming Grove and Woodville townships.

Thirty-two miles of ditch were contracted of which eight miles were completed in 1909. In 1910 the large land dredge operated with two crews working twelve hours each. A smaller boat dredge for working on the branch ditches employed twelve to fifteen men and worked in the northern sections of the township. There were regular items in the newspapers about the project through 1914.

Agriculture–the Major Industry

The township's population peaked at 895 in 1895 and was down to 880 in 1900; by 1910 it had declined to 805. The number of land owners was reduced from 140 to 121, but that did not mean that the township was declining. The number of farmers continued to decline, but they tended to support the firms in the village so it did reasonably well for another half century. Scoville observed, "Some farmers have two full crops on hand… there is more grain in Meriden township than in any other township in the county, but the price does not suit the owners and they are in shape to hold on, and are doing so."

In 1900 the business district consisted of two grocery stores, the blacksmith, the elevator, and the creamery, plus the school. By 1914 the creamery had been greatly upgraded, a cooperative stockyard was in operation, and the elevator was owned by a local cooperative.

In 1901 Charles Pegg and Eugene Tuthill, two of the larger farmers on the north side of the township, shipped three carloads of cattle to Sioux City. Livestock buyers Louis Walther and Anton Schuldt shipped five cars of hogs to Cudahy Packing at Cudahy, Wisconsin. Later in the week Walther shipped three cars of hogs to Chicago. This is a good indication that besides supplying three creameries with milk or cream, the township farmers were generating additional income by diversifying into fattening cattle and hogs. The tax abstract at the time showed that the average assessed value of the land was $16.35 an acre (see the Appendix for sale prices of land), giving the township an assessed value of $415,063 higher than any of the other townships except those with larger villages—Blooming Prairie, Medford, and Owatonna. That yielded $5,844 in taxes in 1904. Much of the success of the farmers came from the fact that the township had 1,311 milk cows, second only to Lemond, which had 1,315 out of 15,536 for the county.

Agriculture was changing. Farmers who had been reluctant in the past because they either did not believe in the "new fangled" scientific ideas or they objected to the professional fees, began to call veterinarians to their farms. In February 1904, Walter Ames, DVM, of Owatonna, was called to the James Andrews farm to give the tuberculin test to the dairy herd. All of the cattle were healthy. In April, J. J. Grass DVM called at the Herman Fette farm. In the years that followed, Grass made many calls, especially to treat horses.

Some of the farmers in the area were caught up by the land boom that took place during the first two decades of the century. In October 1904, L. C. Palas, who had been active in the township, was one of the first to announce that the family was moving to Olivia where they had purchased a farm. That was a comparable farming area, so they would not have to make any great changes. Maria Lutgens and her son Albert made a two week trip to North Dakota and purchased land. They retained that land but returned to Meriden and continued to live there. A.W. Peters and Charles Markmann went on a land hunting trip to Columbia, South Dakota. In the spring of 1905, John Franz, Henry Miller, Joe Walter, and Herman, William, and A.

J. Speckeen went to Hurdsfield, North Dakota, where they met with a land agent and purchased 2,560 acres of land. In May, Juluis Schulke purchased a half section of land in Weld County, Colorado, but it was not until 1911 that he loaded two cars of stock, machinery, and household goods to move there. Ironically, the same issue of the paper reported that L. C. Palas was in Meriden looking for land because he and his family did not like their home in Olivia.

The *Journal* carried ads of railroad excursions for Dakota lands virtually every week. In April 1906, the Fred Bartels family loaded their machinery for their farm at DeSmet, South Dakota. In May 1906, John Franz reported from his new home in North Dakota that land prices were increasing rapidly and land was selling at $3,000 a quarter section. By 1907, Henry Rosenau's brother was living at Grandin, Maria Lutgens was at Goodrich, and H. J. Mundt's daughter was at Barton, all in North Dakota. The Herseys were living in Langford, and the Carl Grunz family had moved to Lemmon, both in South Dakota. At the same time Meriden farmers were leaving for the Dakotas and Colorado, Illinois farmers were coming to Steele County because land was much cheaper here.

The northern Minnesota timber counties had ongoing promotions for cheap land. The Iowa and southern Minnesota farmers who sold out and relocated in that region soon found out that the shorter growing season and longer winters made farming very difficult.

While some farmers were taken in by the land promotions, the great majority in the township were content to work at improving their operations. Horse-powered farming had started in the 1890s, and many labor saving implements were developed. At the same time, steel windmills and gasoline engines were being developed. In 1901, the local blacksmith sold a disc harrow to L. C. Walter and a new steel drag to August Schultz for $25. Gasoline engines were being advertised to be used for pumping water, sawing wood, and grinding feed. Henry Lutgens purchased a thirty-horse-power Russell engine to power his threshing machine. This was not a traction machine, so it had to be pulled by horses.

John Scholljegerdes purchased a mechanically powered saw in 1907 to produce lumber from local trees. Area farmers brought saw logs to his place. Scholljegerdes was one of the more prominent farmers in the area. He wrote a letter on January 30, 1909, to his cousin Eilert, in Germany, in which he revealed the following: He had three daughters who were married and in the township. His oldest son, who was also married, lived on an 810-acre plantation located between Washington, D.C., and Baltimore, which he had purchased in 1908. He made a round trip to visit his son, which had

(SCH) John S. Sr. and Anna Marie Wilker Scholljegerdes c.a. 1902-1904 back row Ida Emila (later Drache) on right Edward i.e. William, middle row left Anna (later Zacharias), Rose center (later Dornquast), front row left Henry, far right Louis. The children identified are those who remained in Steele Contry. Anna married Gottlieb Zacharias and they moved to Wadena, she was the mother of Loretta (Mrs. John Abbe), Edna (Mrs. Lyle Beese), Esther (Mrs. Laverne Abbe), Arthur, Ewald, and Herman who all returned to Meriden.

cost about $96, "but that does not matter to me." He had one daughter in the Alexandria area where he owned three farms totaling 520 acres. In the Meriden/Lemond area he owned 500 acres, half of which was rented out. He also owned 1,950 acres in South Dakota. To operate his 250 acres he had "only eight horses because 100 acres was in forest. But I also keep two horses for going on trips and for milk transportation." They milked twenty-five cows plus had heifers, oxen, and calves totaling about forty or fifty head. In December the milk check was $96 excluding butter (which they received from the creamery). "We had only 49 pigs, [because] the corn last year was too wet. I should have twelve piglets this spring and 200-300 chickens." In 1905, he had thirteen acres of barley that yielded 660 bushels. Wheat yields as high as twenty-eight bushels and corn of sixty bushels per acre were reported.

Throughout years of this chapter the farmers invested in improving their farmsteads and in tiling the land. Some purchased as many as four carloads of tile and others built several buildings. This proved a virtual boon for Hayes Lucas, the carpenters (especially the Korupps), and the tilers. In one week alone Hayes Lucas received seven carloads of tile. At the same time farmers continued to adopt new equipment as it became available. The Klemmer Brothers, who farmed in the Crane Creek community, had a corn shredder and a silo filler and did both jobs at the Charles Pegg farm. This was the first time a silo was referred to in the news. In August 1913, the Owatonna Silo Co. advertised silos for the first time in the *Journal*. Up to this time the standard way to better utilize the corn crop was to shock the corn in the field. During the winter the shocks were hand harvested to get the ears. Then, the stocks were hauled in and fed as fodder. The same was true with threshing small grain. In April 1910, Merritt Scoville spent two days threshing stacked grain, which was the only way smaller farmers could get their grain threshed because those with threshing rigs catered to the farmers with larger acreages. One of the Illinois farmers who came to the area purchased hay and had it baled for shipment to Chicago for the horses used in that city. One week he shipped four carloads and in another week nine. The baler was a stationary, continuous press powered by a gasoline engine. Hay loaders had been invented but were not yet in use in the area, so the loading was still done with a pitch fork.

Probably the first on-the-farm feed grinding took place in October 1910 on the Goodsell farm north of the creamery by George Miller, who had a portable mill. He farmed but did grinding every Wednesday and Friday.

In September 1913, Otto Schueller no doubt was the first in the township to install a lighting system for his farm. The first rural electric line in the nation was erected in Oregon in 1906, but Meriden had to wait a few more years before it obtained that service.

In 1908, under the leadership of Theodore Roosevelt, who was the first president to recognize the need for agricultural reform, a plan of action was laid out. He had endeared himself to the rural people when he encouraged RFD (mentioned earlier) and now, in his words, he wanted "to develop and maintain on our farms a civilization in full harmony with the best American ideals." Roosevelt instigated the movement that led to the Country Life Commission, which conducted a nationwide study on conditions in rural areas. The commission concluded that the two greatest deficiencies in rural life were the lack of organization and proper education. They were followed by soil depletion, health conditions, the plight of rural labor, and conditions of the home. The immediate outcome was the establishment of the extension service and a better roads movement. Cooperatives were recommended to better enable the farmers to compete in a growing industrial world.

The first high school agriculture course was taught in Virginia in 1908, and the idea moved rapidly. In 1910, the OHS instructor was given an extra $10 per month for providing the horse and buggy he needed in his work. That year John Scholljegerdes probably became the first township farmer to travel to Chicago with his son-in-law, Max Drache, to take part in the International Livestock Exposition.

John Scholljegerdes might have been a progressive farmer but his son William (Bill), who was born in 1887, was only allowed to attend the first six grades and then had to stay home and work on the farm until he was twenty-one. Then, he was "sent out to the world supposedly to get a practical education." First, he was sent to Maryland where the family owned a farm operated by his brother. "It was a worn out tobacco farm." He worked there for a few months and then took a job in a cannon shell foundry for thirteen hours a day at $0.12 an hour, earning $21 every two weeks. In the fall of 1909, he went to Elgin, North Dakota. He arrived there with only $8 in his pocket, but after working in harvest fields for three months he had $150. Then, he left for Alexandria, Minnesota, where a sister lived on one of his father's farms, and spent the winter in the woods and worked on the farm for his brother-in-law during the summer. That fall he returned home, where he worked for a "large farmer in Lemond for $30 a month plus room and board." After that he returned to the home farm. On October 11, 1911, he married Marie Krenke, who grew up on a neighboring farm. His father rented him 200 acres of the home farm in Lemond with a house on it that was a converted skimming house that had been pulled to the farm. Bill said, "It was quite large because it required twenty-four horses to pull it to the farm. The house was so cold that water froze in the stove reservoir and in the teakettle on the stove." The farm also had a 34 x 60 foot barn with a stone basement for the cattle. As a boy I helped mix mortar for that. There was also a small chicken coop, a small hog house, and a wood shed that was the original log house. The rent was $162.50, and the farm had 100 acres that could be cropped of which eighty acres were tiled.

Bill had earned five calves when he worked at home, and they had five cows to start his herd. Marie received $1,000 from her parents, and with that money they purchased four horses, one team for $329 and the other for about $275. His machinery line was a sulky plow, a wooden drag, an eight-foot Owatonna seeder, and a used eight-foot Deering binder.

The major crops were fifty acres of small grains, succotash that was used for the chickens and cows, oats for the horses, and fifty acres of corn for the hogs, plus timothy for the horses and red clover for the cows. Bill employed a full-time hired man and another part-time. Scholljegerdes said that a man could average about fifty bushels a day husking corn. He commented that the hogs and the cows were steady income, and the egg money was for groceries. Bill and Marie did all their shopping at Luhmans. They formed a cream route with two neighbors so they only had to go to Meriden once a week. As soon as he could, he started tiling the wet land using five-, six-, or eight-inch tile as needed. Bill recalled, "The Buecksler brothers, Walter, Oscar, Paul, and Art, could lay 500 to 600 feet of tile at two to four feet depth per day. In 1913 so many were tiling that I had to go to Otisco to get tile because Hayes Lucas in Meriden could not keep enough on hand."

Schulljegerdes' routine gives a good idea of the labor intensity of horse-powered farming. Bill Schulljegerdes was my great uncle, and I so enjoyed knowing him; he was always willing to take the time to visit with me. I enjoyed working for him one harvest season. It was my first experience driving a Twin City tractor on a binder. He and Marie were a hardworking and well-organized couple. I never heard them complain about how bad conditions were. The farm work was always done in a timely manner and their farm place was always immaculate.

The government had sent seed to farmers so they could learn what better seeds could do for their production. Apparently that effort paid off, for in 1912 the *Journal* carried its first seed advertisements. The following year American Steel fence posts were advertised for the first time. During the month of October 1913, William Beshman and Carl Brase purchased a corn shredder as did Carl Schroeder and Theodore Pump, and Herman and Ewald Schuldt. In 1914, fifty farmers in the township seeded alfalfa because they had learned the value of improved hay from the newly initiated dairy campaign.

That year the first tractor came to Steele County. It was owned by E. T. Winship, who farmed on the western edge of Owatonna Township and in eastern Meriden. Owatonna Tractor Co. had a machine "that would revolutionize farming." Owatonna was reputed to be the home of the twine binder knotter. It is reputed that it was perfected in the shops of Owatonna Manufacturing Co. by employees H. A. and Watson Holmes, who worked with Appleby who had patented the twine knotter years earlier, but it still was not working to perfection. T. H. Fraser was the local farmer who owned that binder.

Daily Life

The October 17, 1900, *Journal* had two announcements that indicated that the township was at a watershed period in its history. The first announcement stated that the sermon in the German Methodist Evangelical Church would be preached in English. The other announcement read, "About 200 turned out Monday evening for the Republican rally. An excellent address in German was made by Mr. Achillies, and John R. Morley, the candidate for representative, also made an effective address." For many years German remained the spoken language with some families. All four of my grandparents spoke it when addressing my parents, who replied in the same way, but when they spoke to me or my sisters it was always in English. When I started Sunday school half of the class members were equally capable in both languages. It was not until my college years that I regretted that my grandparents had not conversed with me in German. My parents never conversed in German to any one but relatives, and it was always English in our home. I do not recall ever hearing any foreign language spoken on the streets of the village. I understood German, and during the 1950s when we lived near the Monterey Ballroom south of Owatonna, we were still on a party line. My mother would call me and we knew "rubber necks" were on the line, so if she wanted to talk about a private matter she would switch to German and immediately they hung up.

As soon as the crop season was over the two halls in the village started having social events. The Woodmen Lodge was quite active in the village and had many public events, but Grandprey Hall was larger and had most of the activity. In January 1901, a masquerade ball ("masks were on sale") was held at Grandprey Hall, "which was well attended." After the entertainment was over everyone enjoyed an oyster supper. The proceeds went toward the purchase of an organ for the school. The next event was a goat and dog show, which had a "good attendance and was enjoyed by all." It was followed by an event sponsored by Mrs. Sam Grandprey and Miss Edith Patten. It was booked as an "oyster stew social after which a program of exercises were rendered." The proceeds were also for the organ at the school.

Every January, Grandprey held an anniversary ball to commemorate the building of the hall. Grandprey knew how to promote. For example, in January 1901 the three-piece Meriden Orchestra "accompanied by two cornetists from Owatonna" provided the music, and his son-in-law called the square dances. Supper was served at the Meriden Hotel, at which time Mrs. A.W. Peters "was awarded a beautiful water set for being the best waltzer. The event will be long remembered by all present."

The month closed with a home talent presentation of "Under the Laurels," directed by Mrs. Sam Grandprey and Miss Clara Goodsell. Again, the purpose was to raise money for the organ for the school. The ladies were persistent, and finally in October 1904, they prepared "a regular old fashioned supper at the hall consisting of pumpkin pie, pork and beans, etc. for the benefit of paying off the school organ," and they succeeded.

The big event of February by the Meriden Amateur Dramatic Club was "Above the Clouds to a capacity crowd at Grandprey Hall." The receipts were $27, which went to the new Catholic Church in Waseca. A dance followed. "The work done by the members of the club was so good that it would be hard to single out any one for special praises." Later, a home talent group presented "A Bunch of Nonsense" at the hall for a benefit for the Otisco band. The proceeds were $30, "a very pleasing return." During one February a great number of people were ill with the "grippe," which curtailed much of the normal social activity. When an epidemic hit the village it was quite normal for doctors to make calls to the village, and in some cases the schools were closed for two or three weeks.

In 1905, a party of twenty young people took the train to Owatonna to enjoy the production of *Hoity Toity* at the Metropolitan Opera House. When Medora Grandprey, a senior at OHS, had the lead in the class play *The Colonel's Maid,* which was staged at the Metropolitan because it had a larger stage, so many wanted to attend that both evening trains had to stop to accommodate them. Frequently, individuals sponsored dances but charged admission to cover the costs of the hall and the orchestra. For example, Albert Jorgenson arranged and sponsored the May Dance in 1906, and thirty-five couples attended the dance and supper served by the Luhmans. "It was a decided success." I was unable to find out what the hall charge was, but the musicians each usually received a dollar for the night. I do know that in some cases free drinks was the satisfactory reward.

The Meriden Brass Band was organized August 2, 1907, by Herman Schuldt, who became president; Henry Wicklow, secretary; and Peter Pump, treasurer, plus thirteen other members. They were serious about their purpose, and at their second meeting they made arrangements for purchasing instruments. Harry Grandprey was named director. To raise money for their uniforms they held an afternoon and evening bowery dance. "A big bowery 60 x 60 feet was erected and tickets for the dance 75 cents. Supper was served during the evening." They engaged a music instructor from Albert Lea who traveled up on a weekly basis to give them instructions. Six of the members were also members of the Owatonna Imperial Band. In 1912, they played at the state fair for three days.

In addition to dances at the halls, there were many home dances and barn dances. It was common when new barns were built to hold a dance before any hay was put in the loft. Often in the spring when the hay lofts were empty, a farmer held a dance. Clifford Waumett had a dance in what the paper described as "an unusually large barn with a good floor." In one week in June 1909, three barn dances were held in the township—one at Waumett's, one at John Dornquast's, and a third at the August Heinz farm at which the Meriden Brass Band played. In May 1910, the Meriden Brass

Band played for a barn dance at the home of Mrs. Minnie Nelson west of Meriden, and sixty-one tickets were sold. The next night the band played for the birthday party of Mrs. Fred Abbe south of Meriden.

In the winter season dances were held in the homes. The L. C. Wheelocks had thirty couples at their home one mile west of the village on a February evening after which "supper was served at a late hour." When cards were played, the usual games were cinch, flinch, crokinole, chimps, or pinochle. In every case lunch was served after the entertainment.

Mrs. Sam Grandprey "had a delightful social event when twenty-five friends mostly from Owatonna, only six from Meriden, spent the day. Music was rendered by the Meriden Orchestra. A delicious dinner and supper were served." On another occasion she entertained the married women of the village with a "rag bee."

In 1896, the motion picture industry had made its commercial debut with silent film accompanied by piano. On December 30, 1901, it reached Meriden. "A moving picture entertainment at Grandprey Hall took place and the largest phonograph in the world was played between acts, when the reels were changed." It was not until March 2, 1913, that the second moving picture was presented at the Hall. It was presented by the Werdien Brothers, former residents but now from Ottertail County who were visiting their sister, Mrs. Gottlieb Haas.

The motion picture industry was sweeping the country with "movie houses" appearing everywhere. Meriden youth would not be denied. The Meriden correspondent reported that in May 1914, "several youngsters motored to Waseca to attend a moving picture." In December that year another group attended the Gem Theater in Owatonna, where they saw "nothing but the best—the Mutual Novices and Keystone Comedy." Tickets for adults were $0.10, and children under twelve paid $0.05. Matinees were offered four days a week.

Meriden had its softball and baseball teams, just like most of the small rural communities of that era. In August 1903, the Meriden Indians and the Meriden Hoosiers game "was called for darkness with the score four to four. It was the best game played in Meriden this year. Henry Wicklow [then a store clerk] pitched for the Indians and Ray Patten for the Hoosiers. Patten [a Meriden boy] was a pitcher for a Duluth team." In 1906, the Southern Minnesota Baseball league was formed with Owatonna, Austin, Albert Lea, Winona, Faribault, and other teams. But the county townships continued to have their teams. In June 1908, the Meriden North Stars defeated the Meriden Sluggers in a Sunday baseball game. The following week "a large crowd attended the game in Ahlers pasture. The Meriden

Tigers defeated the White Slippers, formerly the North Star team, four to zero." Meriden had a third team, the Meriden Hummers, that played fifteen games and claimed the championship of the league that was made up of Deerfield, Lemond, Meriden, Somerset, and Summit townships. "The 'slab artist' was Joe Pavek striking out 217 in twelve games, an average of eighteen per game. A picked team from the county played against the Hummers and was defeated twenty-five to six. The feature of the game was Young, the Hummers second baseman, who made three home runs in the first three innings, each time with bases loaded. The Pavek brothers were the battery for the Hummers."

In 1913 the township had two teams. One played Otisco at St. Paul Lutheran's annual picnic, after which there was "a grand display of fireworks." Another weekend their opponents were Owatonna Merchants, Blooming Grove, and Steele Center. In April 1914, the team had their picture taken in their new suits. They started their season on May 16 against Beaver Lake, which they defeated fourteen to twelve in ten innings.

Hunting and fishing continued to occupy the interest of many in the area, whether out of necessity, food or bounty, or sport. Ed Bonnell and Charlie Nickolson bagged twenty-five rabbits and squirrels one afternoon in December 1903. Soon after that George Ebeling speared a thirty-five pound raccoon on his farm. Coons were trapped or hunted because they dug holes in the ground that were dangerous to livestock in the form of broken legs. Their pelts also had value. One May eight men spent a day at Clear Lake "and returned home without a single specimen of the finny tribe." On another occasion a group of twelve individuals spent an afternoon at Clear Lake and returned with 200 pounds of fish. L. C. Palas and Rudolph Daetz spent a weekend at Waterville and returned home with seventy-five pounds of fish. Ray Patten had an exciting day when he trapped a mountain lion in the Goose Lake area. It had killed deer, which made it a target for local sportsmen. Merton Parcher, operator of the Meriden Hotel, paid $75 for the hide.

Throughout this period the Meriden Gun Club met regularly wherever they could find a pasture that was not occupied. Herman Stendel broke twenty-three out of twenty-five clay pigeons, but one bright Sunday afternoon at a regular shoot Louis Domy and Sam Grandprey both shot a perfect score of twenty-five. Five fishermen went to Clear Lake by automobile on a Saturday in May 1914 and caught seventy-five perch and forty crappies, but because of a rain storm they had to return to Meriden by train.

Some of the residents were able to travel to distant places. Some even went abroad, and several returned to Germany during these years to visit their families and, often, to help them come to the United States. In December 1903, Mrs. Adam Scheuller went to visit her father in Germany. In May 1904, Mike Ebeling, Sr., spent two months in Germany. When his son Paul finished seeding he left for Colorado to search for land. That year Mrs. Henry (Medora Hersey) Rosenau, and son Ralph, had spent the winter in California for her health. She became worse on the train while returning home and died. Mrs. Rosenau was a sister of Mrs. Sam Grandprey.

In June 1904, Henry Drinken, L. H. Schuldt, and A. Schroeder went to St. Louis to participate in the Louisiana Purchase Exposition. In 1908, Charles Enzenauer went to New York to meet his mother, who was emigrating from Germany. She returned with Charles to make her home on the farm. Every year several groups traveled to the state fair. On a less happy note many people traveled to the Mayo Clinic and the Twin Cities for surgery or for other medical reasons. In September 1909, when six-year-old Otto Drache had a severe attack of appendicitis, his parents took him to St. Mary's in Rochester where he remained for about two weeks.

In 1903, District 48 was only one of five schools in the township and only one of fourteen out of ninety-seven in the county that were open in September because they had nine-month terms. The other eighty-three schools still had seven- or eight-month terms. In 1904, District 48 was closed in the spring for three weeks because some of the students had an infectious disease, which resulted in school going into June.

In 1905, there was discussion because the district had forty-two students, and only one person applied for the teaching position. Apparently the rural schools were not doing the best job because only three students out of the ninety-five eighth graders, had completed their requirements. There were other eighth graders but they were just biding their time until they were sixteen and did not have to continue going to school. Things had not improved much by 1911; only thirty pupils from nineteen districts in the county completed their state requirements. That year four rural districts consolidated with Medford, which was the beginning of the consolidation movement in the county.

In the previous chapter there was discussion about the future of the German Methodist Church in the village and whether the congregation should continue. In September 1903, Rev. Lloyd, of Waseca, conducted services in English on Sunday afternoons and Miss Jackson, superintendent of the Sunday school who lived a mile west of the village, held classes prior to the regular service. In 1905, Easter services were conducted in English at 7:30 p.m. by Rev. Zabel of Morristown. A special program was presented by Ida and Ella Mundt, Esther Keating, Esther Krause, Hulda and Lillian

Peters, and Mrs. O. E. Schueller, all women of the congregation. In 1907, Rev. Zeick, of Waseca, preached his farewell sermon before leaving for St. Charles. He was replaced by Rev. Krienke, of Pipestone, who served both Meriden and Waseca. In September 1913, the German Methodist Evangelical Conference took steps to affiliate with the English Conference. At that point the German group still had nine conferences with 60,000 members. Locally, Owatonna had both churches and Blooming Grove also had a German congregation. In January 1914, Rev. Zeick of Rochester held a revival with services each evening Thursday through Monday. The congregation was still very much alive.

A few news bits of interest from the St. Paul's Lutheran congregation proved that it was thriving. In 1901, the German Parochial school conducted classes four days a week for twenty-two students from five school districts. Each year they went as a group to Owatonna to have their photograph taken.

An oddity took place at the church on September 28, 1910, when Rev. Harrer conducted the wedding of Lizzie Abbe to Ernest Abbe, no relation except for having a German name. After the usual wedding dinner at the bride's parents home, the couple set up housekeeping on the groom's farm, which was less than a half mile from the church. Here they spent the rest of their working career.

In 1912, thirteen-year-old Paul Drache began playing the organ for the St. Paul's services. His parents had purchased an organ in 1910, and the youth took piano lessons from Rev. Harrer, for which the pastor received a load of hay for every six lessons. When Drache attended college at Wartburg from 1919-1921, he returned frequently but substitutes took his place when he did not make it home. While at Wartburg, he took advanced organ training. His parents lived two and one-quarter miles from the church, and in the early years it was his responsibility to get to church whenever music was required. When no one else was going that way he walked, bicycled, or took a buggy or sleigh—whichever method was the most practical.

It was earlier mentioned that the first decade of the twentieth century opened society to the moving picture, which had a significant impact on how people passed their free time, but more was in store for them. During this decade, Americans, more than any other national group, were exposed to the automobile. This innovation not only enhanced the transportation of farm products and the delivery business in every village and city of the nation, it also led to a great, costly, economic, industrial, and society changing love affair.

Like all changes, there was a period of resistance. On October 24, 1903, Mr. & Mrs. Henry Rosenau drove in their buggy (eighteen miles) to Ellendale for Saturday afternoon. Even with good road horses, one way would have taken about three hours on a pleasant late October day. This gives an idea about speed in the horse-and-buggy age. Rosenau liked buggies, and in May 1906 he purchased a new double carriage (a two seater). The same could be said for D. C. Ross, who "was the happy possessor of a new covered buggy."

But signs of change were taking place, and in August 1905 Owatonna experienced its first runaway by a team of horses caused by a car. In September that year, Mr. & Mrs. Louis Fenner, of Owatonna, apparently became the first people to drive a car on the streets of Meriden. In the spring of 1906, the *Journal* reported that the roads "need the attention of split logs." Such roads were called corduroy roads and were very commonplace in heavily wooded areas where timber was expendable. It enabled cars to go over the mud holes. Only two months later, an Owatonna motorist was fined $17.35 "as a demonstrator for an automobile company, for violating the eight-mile-per-hour speed limit. Witnesses testified that he was going ten, fifteen, twenty-five miles per hour." He was fined for the "privilege of using the streets for demonstrating."

In 1907, the first Meriden residents became involved in the Good Roads Association's "lively meeting." The next big news in the township about roads occurred when a threshing rig owned by the Klemmer Brothers of Owatonna fell through a bridge over Crane Creek and was damaged. The Klemmers claimed that the town board knew the bridge was unsafe "but had failed to make any response to the notification given them." The Klemmers offered to settle for $750, but the board did not respond. The article stated that the case could establish a precedent for Minnesota. "If the town board of Meriden is held in this action, it will probably result in fewer people willing to be candidates." The Klemmers sued for $2,300, but the results were never published. However, in June 1914 a short article stated that the Minnesota Supreme Court had ruled that township officers were liable when roads were left in bad condition.

In January 1908, John Ebeling, who was the township's representative on the Steele County Good Roads Association, reported that $2,000 had been approved for twenty-five miles of good roads, all leading out of Owatonna, and another $200 for the Crane Creek road plus $500 for the Lemond road. Meriden Township had not reported any road work except for the bridges over Crane Creek. In 1909, Minnesota passed a law that required that 10 percent of the liquor license money in towns under 10,000 population should go to support roads. This was to be added to other taxes already levied for roads. Steele County Commissioners reported that they were pleased with the "work out" (poll tax) system in Meriden and three other townships for their excellent job of improving the roads. The north Meriden road was improved as a result of the effort by members of the Owatonna Auto Club and also by members of the Waseca Club. "The road is as level as a billiard table." The north Meriden road later became Highway 14 between the two towns. After that project was finished, a picnic was held at what was then the Pegg farm located between sections three and ten.

In 1911 the law required that all vehicles had to have two white front lights and one red light in the rear, "the same that applies to horse drawn rigs." The same issue of the *Journal* noted that Mr. & Mrs. Henry Drinken and Miss Luella Wicklow were driving to Owatonna via a team and buggy and met a car that turned to their side of the road and ran into the buggy. Only Mrs. Drinken remained in the buggy. "The car occupied by three young people was driven by a young girl." Two weeks later the correspondent reported that Gust Fette was driving his horse and buggy and met Alvin Buelow with his car. "Mr. Fette was driving fast [he was always in a hurry] and when he turned out to meet the car the buggy tipped and rolled over on him." Buelow took Fette to a physician in Waseca where it was discovered that he had one leg broken just above the ankle. That probably made Fette the first in the township to be injured in a car accident.

The first record of garages being added to farmsteads came on March 29, 1912, when Ewald Schuldt and Mike Ebeling each purchased lumber to build a garage 30 x 12 x 8 feet for cars. After Ebeling finished his garage,he purchased a five-passenger Buick that cost him about $700 to $800. Meriden did not have an auto repair garage, so in 1912 when Alvin Buelow needed to have repairs done, he drove to Owatonna and returned home by train.

Steele County had the distinction of having the first concrete road in the state in Owatonna, but Meriden was not left out on road improvement. The road was straightened to aid traffic between Owatonna and Waseca. The biggest news in 1912 about roads was when it was announced that the proposed east-west route from Winona to Mankato received a federal grant of $10,000 to improve a fifty-mile stretch of post roads. Local news was that one night George Joriman, of Meriden, was arrested for excessive speed. He whipped his horses into "a mad runaway from Oak Street west [in Owatonna] and made no attempt to control them." His buggy was smashed but he pled not guilty.

In May 1914, the Meriden town board traded their road grader in for a new heavier model. Two firms were each given 2,300 feet of road to properly grade. About a hundred spectators plus the boards of Meriden and Owatonna townships observed the demonstration in section thirty-six. The township purchased a new grade and then secured a lot from Clara Goodsell for a machine shed. A building was erected on that lot next to where Grandprey Hall would be moved in 1915.

In the summer of 1914, two traffic events happened with different outcomes. Martin Ahlers drove past the John Ebeling farm at "a pretty fast pace going over a narrow grade and got off the crown of the road." He was accompanied by George Enzenauer and a Nelson boy. He hit a wide uncovered culvert that crushed the front wheel. He received a broken jaw, lacerated arm, internal injuries, and was paralyzed from the waist down. He died fourteen days after the accident. Ahlers was survived by his parents Mr. & Mrs. John Ahlers; brothers Herman, William, Albert, and John; and sisters Mrs. John Wilker and Mrs. Albert Wilker. This was probably the first auto fatality of a Meriden resident.

An accident with a team and buggy within in a month of the above occurance had a far different result. Lloyd Grandprey and five friends were returning to Meriden from Owatonna when the team was frightened by a dog. That caused a bolt to slip out of the buggy tongue, which then broke. The buggy went into the ditch but no one was injured. "The team ran home nine miles in fifty minutes right up to the barn door." Lloyd walked to the Bartsch farm on the west edge of Owatonna to call Wicklow store. His father, Sam, had just walked into the store and heard the conversation, so Sam knew what happened. Lloyd borrowed a team from Bartsch who helped fix the buggy, and the group "all returned to Meriden by 1:00 a.m. tired but fine."

The following are miscellaneous events that happened during these years. F. W. Goodsell (mentioned previously) was one of the pioneers of the township, having filed for seventy acres in the northeast quarter of section eighteen north of the railroad. In years that followed his widow, Clara, secured the north 160 acres of the section and lots in the village in section seventeen south of the tracks. Later she sold lots to people who wanted to settle on the west side of the county road north of the tracks, which included the creamery lots. In October 1903, her daughter-in-law, Mrs. E. E. Goodsell, and daughter, Clara, left Meriden for their new home in Waseca. The pioneer family of the village was no longer represented there.

A major scandal occurred in June 1907 when Otto Schueller was accused of raping Emma Neubauer, who was a domestic in his employ and "a prisoner in his farm home." In January 1908, Schueller was acquitted of rape charges "with his hired girl Emma Neubauer" but then was sued for $15,000 by a Mr. Karsten of Morristown, the girl's grandfather. No further articles were found about the case, and the Schueller and the Karsten families continued to live in the area.

One of the last signs of pioneering took place in the spring of 1909 when a large body of immigrants "with several nice covered wagons and other conveyances camped in the village overnight because of bad roads." The roads must have been extremely bad because even the local dances were cancelled.

A major event took place in Owatonna in February 1910 when C. I. Buxton was engaged as the secretary of the Minnesota Implement Dealers Mutual Insurance Co. and brought the headquarters of the company to the city. Later, it was reorganized as the Federated Mutual Insurance Company, and many area young men and women had an opportunity to work there after they finished their education.

Meriden got its first sidewalks in 1898. They apparently were not satisfactory, and in 1912 many businessmen decided they needed to redo and extend them. The new walk started in front of the post office, went past the blacksmith shop and Wicklow store, and in front of the next three homes to the Markmann residence, which later housed the post office.

The local businesses went all out to promote the community when on September 8, 1913, the Frank T. Collins "Village Circus under tent" appeared in the village. The correspondent wrote, "The best entertainment of its kind that has ever visited Meriden. The large crowd who attended was highly pleased." Discussion took place in the weeks that followed about attempting to have a circus the following year, but apparently the sponsoring businesses were not impressed with the results.

H. C. Bartles erected a building on the south edge of the village on May 20, 1914, and opened a "wholesale" liquor store. The only other reference to his venture was in a later issue of the paper, which stated that on June 1 he had installed a Northwestern Bell telephone. The word "wholesale" might have meant "off sale."

Journal; Casey & Douglas 217; Minutes of the Meriden Rural Telephone Co. 1913–1961, in the files of the Steele County Historical Society, hereafter Meriden RTC; *Chronological Landmarks* 35-42; Clayton Ellsworth, "Theodore Roosevelt's Country Life Commission," *Agricultural History Journal* XXXIV (1960): 155, 156, 160, 166, 167, 170; William E. Scholljegerdes, personal interview, 7 July 1979, hereafter Scholljegerdes; Drache, *Challenge* 88–89; Robert C. Davis, "The Impact of Mass Communication," *Technology in Western Civilization*, vol. II, 326.

CHAPTER V
The Village Reaches Maturity
1915–1929

Thorstein Veblen, one of the nation's great economists of the late 1800s and the early 1900s, stated, "The country town, even though a major and enduring part of our culture was one of the most wasteful institutions to ever exist. Most were capable of serving far more farmer customers than they ever had, so to survive the small town merchant had to charge exorbitant prices or go broke. A few big profits were made, but many more small ones were lost." The reason many merchants failed was that many of them allowed customers to use credit, and if a customer went broke, or for another reason decided to leave the area, most of the time he or she did not pay the grocery bill. That appeared to be less of a problem in areas where settlers had to purchase their land than where they acquired land under the Homestead Act.

In the previous chapter it was stated that both the township and the village had already experienced a decline in population, but based on what has been written thus far no land speculators or merchants had made any great profit in the first sixty years. It appears that there were only two people who made any significant profit from the development of Meriden: F. W. Goodsell and L. C. Palas. Neither of the two were the original buyers of their land when it was first sold in 1856, but Goodsell made a profit because he had purchased about seventy acres of land in section eighteen that was north of the tracks and west of the county road that ran through the village. He sold eleven lots from that property. Later, family members purchased other parcels along the railroad on the south side of the street to the east and another 176 acres in the north half of section eighteen, which Clara sold to Louis H. Walthers for $12, 500 ($71/acre) in 1925 when better land was selling for $125 an acre. Mrs. E. E. Goodsell, the widow of F.W Goodsell's Son, and her daughter Clara, lived in the village until 1925 when they moved to Waseca and built a new house. At that time they still possessed the farm north of the creamery and west out of the village. L. Peter owned the seventy-two acres in the northwest quarter of section seventeen, which was north of the tracks and east of the county road. Palas did not secure his land until after 1900 when the early pioneering days for the village were well past. He sold some lots including two acres to the school district and was an active developer in the community. He sold windmills and for years had threshing and corn shredding rigs.

My parents became active in the village during this period and were in business long enough to see it mature and decline. By 1915 all of those who came to the village to speculate had come and gone except Henry Wicklow and H. J. Luhman who had the most enduring privately owned businesses. Hayes Lucas Lumber, Company was the longest lasting business owned by an outside entity. Its business was hurt when the farmers created a cooperative elevator because it sold coal and tile, which were major items for Hayes Lucas as was salt for livestock, which was also handled by the general stores. The elevator also sold flour.

I remember the day I was with my dad at the store when Wicklow told us, "If I can't make twenty-five cents on every dollar's worth of goods I sold I would quit." He and Mrs. Wicklow worked full time to make the business succeed. He took only one vacation during his working years, and Mrs. Wicklow took two short ones, one to Chicago and another to South Dakota. Veblen understood the situation, but how else could the frontier have been opened without the small farm service centers, considering the transportation system of those days? In addition to Wicklow, Luhman, and Hayes Lucas, there were only about a half dozen families who acquired more than what was needed for a comfortable living.

The Chicago & Northwestern

In 1915 the village was still served by three passenger/mail trains going in each direction daily. The schedule was good enough so sisters Esther and Bonita Abbe could board the train at 7:22 a.m. for school in Owatonna and return to Meriden at 5:30 after classes were over. They preferred this rather than rooming in Owatonna, and with the round trip commuter rate of $0.23 that Virena Stendel paid, it was probably more economical. When she stayed in town she paid $4 a month for room and board, but she also brought some food from home and had to help with house work.

The depot in 1919 when repairs were being made on the platform. Note the field cultivator parts on the platform and the section car. The Chicago Northwestern Railroad Historical Society photo.

In 1916 a new, fast train between Mankato and Chicago meant there were four passenger trains, but this one carried no mail and made only limited stops. The trip took fourteen and one-half hours to make the daily run. In anticipation for this heavier, faster train, the tracks from Mankato to the junction near Winona had to be upgraded. In January 1917, all train service was stopped for two days because of heavy snowfall.

Steam engines were notorious for causing fires, and Meriden was not immune from the threat. In October 1917, Gus Schendel, a bachelor who lived south of the tracks on the "east end," saw that there was a fire at the lumber yard soon after a train had passed through town. He woke others and they broke into the hardware store to get pails. Fortunately, Hayes Lucas had water barrels at each corner of the lumber shed, which was ample to put out the fire. Manager Domy was out of town. Not long after that event two trains collided at the end of the siding on the east side of the village. Two empty freight cars were wrecked and both engines were damage.

There were other inconveniences such as in December 1919 when, due to a severe national coal shortage, the road was forced to reduce service to two trains each direction instead of four, but within a week normal service was resumed. In August 1920, an axel broke on a car loaded with crushed rock and jumped the track. The caboose was tipped in the process, causing injuries to the conductor and slight bruises to the brakeman. The track was blocked from 10:15 p.m. to 4:00 a.m. In the winter of 1926, the CNWR advertised special fares for the International Livestock Exposition and Horse Show at the Chicago Union Stockyard. This was the premier livestock event of the nation from 1899 until the 1950s when the stockyard was moved out of Chicago to Peoria.

The 1925 section crew and hand car. L-R: "Rollie" Mueller, Louie Werdien, Bill Mueller.

Between 1916 and 1920, American railroads had a building surge and peaked out at 254,000 miles of track. The roads quickly learned that they had over expanded and almost at once started to abandon some tracks. Most of the railroads experienced a decline in passenger traffic, and the CNWR was no exception. In December 1928, it appealed to remove one eastbound and one westbound passenger train between Winona and Tracy due to the small amount of passenger traffic. It also requested permission to close the Haverhill station east of Rochester because "business at that station did not justify keeping it open." By that date the road had already discontinued the passenger train between Tracy and Watertown, South Dakota, and "substituted it with a gasoline driven one-or two-car train commonly known as the Galloping Goose." Even special round trip fares of $1.75 to the University Farm in St. Paul for the Boys and Girls 4-H Club Excursion could not keep people from using the automobile. The American love affair with the automobile had started, and it would have a major impact on small towns like Meriden.

Meriden Businesses

Meriden Rural Telephone Company

Even though some people voiced the traditional opposition to the telephone, once they realized how great it was not to be isolated, they appreciated its usefulness. All but the most rugged individualist wanted to be connected to the outside world. The MTRC held its first annual meeting in 1914 in the courthouse, after which meetings were held in the District 43 schoolhouse because it was near where several of the officers lived. In 1917, an assessment of $3.00 was levied on all members. The phone of the L. G. Campbell Co., was transferred to the Meriden Farmers Elevator Co. which was now owned by the local cooperative. In 1919, the Waseca Telephone Co. extended service to four farms northeast of Meriden. Those farmers did business in Waseca and had relatives who lived in the Waseca area. The village already had Owatonna and Waseca phones because the businesses had customers in both directions. In 1919, Steele County had twenty-three rural telephone lines with about 800 subscribers (an average of thirty-four subscribers per line), and they were all members of the Tri-State Telephone Co., which provided the long distance service. Tri-State charged $3 per subscriber per year for the service, paid in advance. In July the government ordered an increase to $6 per year and ordered that all rural lines that did not pay would lose their service. The association members passed a unanimous "no" vote because they felt that most of the difficulties were within the limits of Owatonna and not in the rural areas. It was later realized that the move was a major political blunder because the problem for the poor service was difficulty in securing adequate labor.

In 1920, it was voted to charge renters $15 per year, and members were assessed $5 annually. The secretary and treasurer were each paid $2.50 per year for their duties. The lineman was instructed to purchase a safety belt, and Wm. Burr, the lineman, was paid $0.60 an hour. Everyone, renters or members, was assessed $2 when a telephone was installed or moved from one floor to another. At the 1924 meeting it was moved that renters had to pay their $15 by February 1 or their phone would be disconnected and removed.

Business was increasing enough so another lineman had to be added. Burr was assigned the 730 and 905 lines in the eastern area of the township, and Andrew Matthew was assigned the 723 line, which included the village. More independent lines were established as the number of subscribers increased and as it became more difficult to make calls when the phones were used for entertainment as well as business. For example, The Independent

Rural Telephone Co. was organized in 1928 by about a dozen families, all members of the Worthwhile Club who resided in a couple sections along the Meriden/Owatonna township border. A storm in October 1925 was so intense that it caused the lines to be out of order for two weeks. In 1926, it was decided that the MRTC should carry liability insurance. That proved to be a wise decision.

Post Office

On April 1, 1917, Samuel E. Grandprey resigned his position as postmaster and announced that they would be moving to Owatonna to operate the Alpha Hotel. Gladys Rosenau became acting postmaster while Civil Service examinations were conducted. Henry J. Luhman was named postmaster, and in September 1917 the post office was moved to the Luhman store. After a decade of complaints about the service by community members, Luhman was given notice that there would be a change, which was not to his liking. On July 7, 1928, Paul and Anna Drache took the Civil Service examination. Paul scored 76.10 while Anna scored 78, and on December 3 Anna was named postmaster; Paul was named assistant on December 8th. On December 28, 1928, the inspector in charge wrote to Postmaster H. J. Luhman, "You have refused to turn over the office.... It is not believed you will experience any difficulty in making the transfer."

On January 1, 1929, Anna Drache started a career that lasted forty years. Her first salary was $20 a month and $16 rent for the 110 square feet of office space and heat and light. Paul received $10 a month for meeting the six trains daily. Anna paid $175 to purchase the boxes and counter, which were moved into the Drache home. (The post office as she had it is now in the Village of Yesteryear.) The first year went smoothly, but the office was in the Drache home and they were advised to keep the office open as long as the two stores were open, which on Saturday, could be as late as 10:00 p.m. Then regulations were changed and the Draches were advised that the office was to close at 6:00 p.m., but Paul felt that postal patrons should be accommodated whenever they came. This was strictly against postal regulations and over the years it caused a few arguments. I remember many times people came to the back door of our home while the family was eating to ask for their mail. A notice in the December 20, 1929, *Journal-Chronicle* read, "The Meriden Post Office to be open all Day Sunday December 22 for the convenience of the patrons."

General Merchants

By 1915, Luhman and Wicklow were well established. Each had a solid customer base and maintained a good business until they were ready to retire. The Meriden business people and some farmers tended to split their business between the stores. There was virtually no comment about the competition. The June 11, 1915, issue of the *Journal* had an article about Wicklow discovering "a good sized tarantula while removing a banana from a bunch. It is on exhibit in a jar at the store." In 1919, Luhman removed the original board walk in front of his store and replaced it with a cement platform and sidewalk in front of the store and the post office.

The village west of the country road in 1916. L-R: German Methodist Church, horse barn, Luhman store; note second stairway for apartments, flour and salt shed, and new livery barn.

It was a big day for everyone in November 1922 when the electric power line arrived in the village. Wicklow had just upgraded his large ice box cooler, so he waited until 1927 to purchase an electric refrigerator and ice cream cabinet.

For many years Herman Schuldt, who farmed a half mile north of the village, had supplied milk on a regular basis to residents of the village. But in 1927, he increased his herd and decided to separate all of his milk, so he announced that he would discontinue delivering milk. Fred Fette, the banker, had two cows, a pig for fattening, and a few chickens kept in the barn next to his house, as did Ewald Fette who had the small farm on the east end of Mill Street. So did Rudolph Schendel, but they were not interested in the milk route. Schuldt asked Paul Drache if he would be interested. The Drache place had a barn and two acres, so he purchased three cows and delivered milk on a daily basis. When he started in 1928 he printed tickets and priced the milk at eight cents a quart. The ticket was perforated so it could be ripped if only a pint was desired. He carried the milk to his customers in covered two-quart pails. I still have one.

The Draches always provided room and board for two teachers and two others. After Anna became postmaster and Paul became involved in other activities, they employed local farm girls to help. Most of these young women knew how to milk cows and took over delivering the milk if no one else was available. Then, Drache purchased pint and quart bottles and a bottle carrier. The customers set out the cleaned empty bottles with the green tickets for the amount of milk they wanted.

In June 1928, Wicklow installed two underground storage tanks and two hand operated gasoline pumps (Hi test and regular) and became a dealer for Central Cooperative Oil Association. I believe the Fette Garage was the first with such pumps.

Hayes Lucas

In 1916, the company added a 30 x 40 foot addition to the south side of the main store, which was used for paint, glass, ropes, and a glass cutting room. They stocked large panes of glass and used a hand cutter to make the size panes requested by the customers. They had many sizes of rope to fill every need on the farm. Most of the paint was in one- and five-gallon cans. In 1919, the company sent James Hennesey to oversee construction of the large lumber shed that filled the space between the coal and cement sheds located along the tracks and the main building. This building had carloads of all dimensions of lumber piled on two levels. This made one of the best hide-and-seek places for the town youngsters. There were several scarred knees and a few slivers gained during the games but no broken bones. Employees always warned people that the yard was not a playground, but the children had a good time as long as they did not get caught.

In 1925, Lloyd Grandprey, who had received his training in yards in South Dakota, became the manager. He was well liked, managed a good business, and the yard and hardware drew customers from a large area around Meriden. Mrs. Grandprey was the former Eva Ochs who had taught school in Meriden for four years prior to their marriage in 1920. The company owned one of the finest houses in Meriden, which it leased to the Grandpreys for $15 per month.

First State Bank of Meriden

The impact of World War I on the local economy can be partially understood by observing the impact it had on the use and the devaluing of the dollar. In 1910, Owatonna had three banks with resources of $1,162,094. There were three other banks in the county, two in Blooming Prairie and one in Ellendale. At that time the consumer price index was 114. By the same period in 1914, the three Owatonna banks had resources of $2,444,482, the two Blooming Prairie Banks had $861,930, the Ellendale bank had $200,088, and the new Medford bank had $61,212, for a county total of $3,566,715. The consumer price index was about 164—meaning prices had increased that much in the previous four years.

The township liked the idea of having a local bank. In December 1914 the *Journal* reported, "Business men and farmers considering starting a bank in the village. Twenty have already subscribed and applied for a charter. A public meeting [is] to be held in January." The January 29, 1915, issue of the *Journal* stated that the new State Bank "in" Meriden would be capitalized at $10,000, and a modern brick building would be erected as soon as weather permitted. There were thirty-eight charter stockholders; no stockholder had more than six shares. Only two of the original stockholders were not from the Meriden trade area: Sid. W. Kinyon and C. J. Kinyon, from Owatonna, were involved with several financial institutions. The site chosen for the bank was where the Grandprey blacksmith shop/hall was located. Grandprey sold the lot for $600 and moved the building to just south of the creamery. The articles of incorporation were signed by the thirty-eight stockholders on January 14, 1915, and charter No. 1159 was issued by the state on March 11, 1915. J. H. C. Schuldt was elected president and Sid. W. Kinyon vice-president. Fred Fette, who came from Germany by himself at age sixteen and apparently received his education before coming to Meriden in 1910 to farm, was named cashier. In addition to the three above, C. H. Wilker, H. J. Luhman, Frank Domy, and Peter Moe made up the original board of directors. E. T. Sommerstadt, of Waseca, was contracted to erect a 24 x 42 foot red brick building with grey stone trimmings for $2,450. At that time the largest contributor had put in $15,000. I suspect that was Sid Kinyon.

In addition to the property and $10,000 capital, the bank had a $2,000 in surplus when it opened Saturday, July 10. The bank stayed open the first evening, at which time it received $7,000 in deposits. By the following Monday evening, deposits totaled $31,000. Fette was assisted at the opening by Walter Finch, an experienced banker, of Owatonna. Soon John L. Westrom, a stockholder, became his assistant. On July 1, 1916, Lloyd

Grandprey, also an original stockholder, resigned his position with Hayes Lucas to become the assistant cashier.

The First State Bank of Meriden, c.a. 1918, Wicklow store on the left and post office on right.
Kathy Schroeder Wett photo.

By December 1917, the resources were $134,108.79 and business had increased so much that the six-digit adding machine was no longer adequate and a ten-digit machine was purchased. (That machine is now in the post office of the Village of Yesteryear.) By this time Fred Fette, who was also secretary of the elevator, the creamery, and the stockyard, needed extra help. In June 1917, Paul Drache, a graduate of Canfield Academy, was called on to assist. His primary job was to write out checks issued by the three firms. His penmanship course proved its worth. He was paid $10 per month.

When Lloyd Grandprey was drafted in May 1918, Drache was promoted to assistant cashier. Business continued to grow, and in September Lenora Abbe was employed. In 1919, the bank also assumed the duty of collecting real estate and personal property tax for the county. When Drache left to attend Wartburg College in September 1919, his salary was $25 per month. In November 1919, the sixty-seven stockholders voted to raise the capital stock to $25,000 and the surplus to $5,000.

In 1920, twenty-five safety deposit boxes were added as a further service to the bank patrons. By July 1921, the bank had made a total of 660 loans and had experienced a profit in all six years of operation. That year it agreed to pay the creamery and shipping association 2 percent on their daily balance. It also installed an alarm system, which was connected to the Luhman and Wicklow stores. In 1923, an upgraded alarm system was installed by Bankers Electrical Protection at a cost of $765 to be paid four months after Meriden received electricity. The elevator was informed that if it put its security in proper shape the interest on its loan would be reduced from 7 to 6 percent. This prompted the cooperative to incorporate. In 1924, Ewald Fette was employed as assistant cashier at $100 per month. His father's salary was $200. By the end of 1924, a total of 2,244 loans had been approved, and the board moved to pay a 10 percent dividend to add $1,000 to the surplus and the remainder to undivided profits. Business continued to grow rapidly, and it was decided that it should have a correspondent bank. By September 1926, it had approved 3,110 loans and opened accounts with the Continental and the Commercial National Bank of Chicago.

Within a month the Kinyon Investment Co. was notified that it should sell its real estate within thirty days or the loan committee would sell it at the best price it could get. Within days, S. W. Kinyon resigned as director and vice-president and sold his nine shares to other stockholders. The year ended with notification to a Williston, North Dakota, bank to send its statement on real estate. The bank then charged off $1,500 but was still able to pay an 8 percent dividend for the year. In 1927, there was still concern about some of the Dakota real estate, so H. J. Karsten and Fred Fette made a trip there to look at the land and the crops.

In 1928, nine banks of the Steele County Bankers Association announced that there would be a fifty cent a month charge on all checking accounts with balances under $50. Throughout 1928 it took over property in Minnesota because of default and moved to sell all land in North Dakota and Montana plus the bonds in the Great Eastern Elevator Co. in Montana. At the annual meeting in 1929 the by-laws were amended so the highest indebtedness "shall be $1 million." The meeting closed with a motion to declare a 10 percent dividend.

Blacksmith Shop, Garages, and Standard Oil

The automobile made its impact on the village business district during this era. L. C. Palas owned the seventy-two acres in the west half of the northwest quarter of section seventeen, which included the land north of the tracks and east of the county road in the village proper. In March 1917, he announced that he would build a 40 x 70 foot garage in the southwest corner of that property bordering the tracks. He stated that it would "be equipped to do all repairs and overhauls of autos." There was no further notice about what happened until 1919 when it was reported that he would operate the garage. He was involved in other agricultural activities, and because of the strong farm economy was kept busy with those lines so the garage remained a sideline. His son, Harvey, who had gained experience at the Sander Garage in Owatonna, took over the garage business in about 1925 and maintained it until he retired. Paul Buecksler, who had an early version of the Model T, stated that he always went to Palas for repairs.

In June 1917, Gustav Fette purchased a "motor truck," which was probably the first one purchased in the village trade area. At the same time Fred Fette (no relation) purchased the lot and the building in which the post office was formerly located but made no comment about his intentions.

The *Journal* regularly had ads about the motor truck and electricity becoming necessities on farms. "The truck will reduce the cost of transportation by 50 percent and about 150,000 farms nationwide already had electric power provided by Delco systems." A Delco ad stated that farmers who figured the time value of their labor knew the ad was true, plus the good income during these years encouraged them to purchase trucks and electrical systems when they were available.

In 1920, a new school was being planned, and the town board decided that they needed a permanent place for a town hall, so they purchased the hall and blacksmith shop from Sam Grandprey for $2,000. That building remained the town hall for the next fifty years. The ground floor remained the blacksmith shop during those decades.

In 1919, the village experienced a boom when four houses were built and Standard Oil Co. constructed a bulk storage facility in the area directly south of the depot. Tank cars of gasoline and kerosene made their appearance on the siding for the first time. In March 1922, Ludwig Schultz, a son-in-law of Theodore Pump and a son of William R. Schultz of Lemond, who was living in the east end of Meriden, was employed to deliver oil and gasoline from the bulk station. At the same time Central Cooperative Oil Association was being organized in Owatonna and did such a good job in getting stockholders in the township that Standard Oil did not have much success.

There was little news about the Standard business until 1926, when a crew came to the village to take down the large bulk tanks and shipped one to Mankato and the other to Owatonna. In February 1928, H. W. Kreutzer, the local Standard dealer, stated that the building would be moved to Waseca. In May the cement piers that held the bulk tanks were removed, and the grounds were leveled. By then Central Coop had established routes in the township and had 1,100 members with a total of $232,100 in sales and was "the largest of its kind in the state." Its 1927 International truck was a familiar site on the township roads. One very cold day in 1932, I had the pleasure of riding in its unheated cab with the delivery man, Elmer Hammann, to the Ewald Wilker farm, where he used five-gallon cans to fill the fifty-gallon drums of gasoline.

1924 view from the elevator looking east toward the Standard Oil building and two bulk tanks, the Hayes Lucas L-R coal, cement, lumber sheds, the hardware store, glass, paint, rope, and harness annex; the stockyard, water tower, and office. The white house in the background was where the Hayes Lucas manager lived. The large house on the right was the home of Fred Fette; the smaller white house was built by E. L. Scoville, and later became the home of Gustav Schendle. This picture was taken by Paul Drache.

In 1923, Fred Fette built a garage for $600 on the lot from which the schoolhouse had been removed adjacent to the county road. His sons, Herb and Ernie, were very mechanical and creative, especially when it involved electronics. Herb's wife, Mildred (Millie), said that they had so much business that the garage was paid for in two years. Herb was only sixteen in 1923, but he had already worked in the Lewer garage in Waseca. Millie stated that when they were first married Herb operated a generator powered with a Model T engine four hours each day to charge batteries to supply power for the house. She said that they probably did not have the money to wire the garage for electricity so they did not connect to the power line.

Looking north from the elevator, on the west side of the country road is the Harry Grandprey house, township shop, blacksmith/town hall, the creamery, the Gontarek farmstead. On the east side of the road is the Harvey Palas house, L. C. Palas house, Papke house, the creamery house, and the new school. This picture was taken by Paul Drache in 1924.

Dick Rietforts, a professionally trained blacksmith, arrived in Waseca in April 1922. His Uncle George had sent him money for his ocean voyage, which was $175; train fare to Waseca was $50. He worked for his uncle for a year and one-half to repay him. Then he worked for a blacksmith in Waseca who had a large business because "everything in town was still delivered by horse and wagon. There was always someone waiting with his horses when he got to the shop at seven." Then he received a call from Ernest Schroeder, a Meriden farmer, informing him that Meitzner was leaving Meriden.

The March 16, 1928, *Journal-Chronicle* reported that Rietforts had completely remodeled the blacksmith shop formerly occupied by Julius Meitzner, who went farming in the Rice Lake area. Rietforts had gone to business school after which he learned the blacksmith trade from a master blacksmith, and during WWI he was a blacksmith in the German army. He also learned wood working in Germany because most of the implements there were still made out of wood. The article stated that Rietforts would do horse shoeing, plow work, wagon repairing, disc sharpening, and other general blacksmithing. Rietforts had a combination machine that planed, sawed, beveled, and performed several other operations, so in his first years he did a great deal of working with wood. He explained how wagon wheels evolved from all wood with a steel tire to wooden wheels modified to have rubber tires. Finally, rubber-tired grain tanks and then trucks came.

Later in life Rietforts was asked why he went to a little place like Meriden. He replied, "It was a lively little town, two garages, a big elevator, two big grocery stores, and a big creamery. So there was lots of stuff going on there." He paid $600 to Meitzner for all the equipment and within a year or two he had money to equip the shop with electric motors. As soon as he opened he said he "had a lot more work than I could do because the farmers did not buy much new equipment so I got to do lots of repairing. I had all the work I wanted. The highest I charged was fifty cents an hour." He continued that everything changed and explained the evolution caused by the automobile and good roads.

His business area extended north half way to Morristown, south past Lemond, and west nearly to New Richland. During the winter he often had 500 or more plowshares to sharpen, so he was able to work every day. His wife, Gertrude, interjected, "I think he was famous for doing good work on plowshares because farmers came from quite a ways." That, too, changed with throw away plowshares.

When Dick first rented the shop the town board charged him $8.50 a month. He had one responsibility and that was whenever there was an event in the hall above, he had to make sure that he put the extra supporting posts in place to make sure the floor would not give way.

In February 1928, Albert Born, a dealer for the J. R. Watkins Company in the Good Thunder area, was transferred to the Meriden territory. Born had been a well driller and was drafted during WWI. After his service he returned to drilling wells until his banker encouraged him to buy land and go farming. He bought land, drilled a well, and started a farmstead. Someone offered him $300 an acre, but he turned that down. He wanted to farm but continued doing well work. While working on a windmill he lost an eye and damaged some fingers so he could not do that work and was unable to farm. So he rented out the farm and with the downturn in the agricultural economy, the renter sold the crop and did not pay the rent, so Born lost the farm. That is when he took a job as a Watkins dealer.

There was no housing available in Meriden, so he had to leave his family of four in Good Thunder. He came to Meriden with a 1926 Dodge panel truck that, besides having Watkins products, also had a tool box. He joined Rietforts at the Paul Drache home for room and board along with two teachers who were also there. In addition to the Watkins route he sold tools and fixed pumps. When the Meitzner family left town to go farming, Born rented the Peter Pump house they had occupied on the north end of the village.

Albert Born, Millie (Anderson) Fette, and Dick Rietforts all came in 1928 at a key time in the community, and each made a positive contribution to the village.

Meriden Farmers Cooperative Company

In 1915, the entire cooperative movement was expanding rapidly, and by , Meriden, Medford, Pratt-Havana, Hope, and Owatonna had livestock shipping associations. The American Farm Bureau Federation was organized in 1920 and as part of that movement the Central Livestock Commission Association was established in So. St. Paul. Central immediately sought to get the local associations to unite with them. The Owatonna association had already joined and encouraged the others in the county to follow. At this time Minnesota had 655 associations that shipped 38,080 carloads, 61.4 percent of sales in the state, but by 1925, 125 were already out of business. A study by the extension staff indicated that failure was caused by poor management, farmers trucking their livestock directly to market, competition of independent buyers, packer buying, and the associations trucking directly to packers. The Meriden association did not join the Central , preferring to ship to packers in Sioux City, Iowa; Chicago, Illinois; Cudahy, Wisconsin; Austin, Albert Lea, and Winona in Minnesota; or to other commission firms in So. St. Paul.

During some weeks in the years 1916 to 1924, as many as nine carloads of livestock were shipped. From then on shipments declined steadily. In the previous chapter the difficulty farmers had in getting their livestock to the local stockyard was explained. Paul Buecksler said that most of the time they drove the hogs along the road. Norbert Abbe recalled that when a bull was brought to the yards one man held the staff to the ring in the nose. A second person walked behind the bull holding two ropes, one tied to each of the front legs, and if the bull tried to charge he tripped him by pulling on the ropes. Cattle were driven by riders on horseback. My home was located along the road to the stockyard, and I remember a few days when the road was lined with animals in all sorts of horse-drawn rigs, often cattle and hogs afoot, but no trucks.

Midwestern agribusiness was still developing, and the So. St. Paul stockyard was still increasing its volume. In 1926, it set a new record of 88,279 carloads and, as the *Journal* stated, 59,993 loads were "delivered by automobile truck." In 1925, the Chicago stockyard had peaked with up to 100 trains of livestock a day, but train shipments declined as truck deliveries increased.

In 1929, the Meriden yard shipped only 42 cars compared to 197 for Blooming Prairie, 215 for Hope, and 417 for Owatonna. Then, in the fall of that year, Theodore Pump, manager of the yard, realized that he had two cows and a bull belonging to Ernest Abbe in the yard and was not able to get a train car. He called Paul Drache, who had a used 1925 Reo Speed Wagon truck, and asked him to haul them to So. St. Paul. Pump rode along. The shipping rate was fifteen cents a hundred pounds for which Drache received $6.30. This was the first truckload of livestock taken to any market from Meriden. It was the beginning of Drache's livestock hauling business and the finishing blow to the Meriden Farmers Cooperative Company. My dad and Lloyd Grandprey said that to their knowledge the association came to an end with that shipment.

There was never another article in the *Journal* about the yard and no records were available. In 1930 or 1931, Donald Born and I went into the yard office (a little yellow building) and read many of the scale/sales tickets that were scattered on the floor but did not save any of them. It is my impression that the water tower and the stockyard were dismantled by the railroad in 1932.

Meriden Farmers Elevator

In September 1916, the L. G. Campbell Co. offered to sell its Meriden elevator. The local farmers met soon after that announcement to determine if they wanted to form another cooperative to operate the elevator. At the second meeting twenty-six out of twenty-nine voted to organize the Meriden Farmers Elevator Co. Seventy-nine shares at $10 each were subscribed for with no one buying less than two shares. They voted to capitalize at $10,000, and as soon as they could raise $3,200, they would submit their offer to buy and then make a lease with the CNWR. Fred Fette chaired the meeting and was also the major spokesman for the group. He stated that at least 150 of the 250 stockholders in the shipping association would subscribe for stock in the elevator association. J. C. Wilker was elected president; Herman Schuldt, secretary; and J. H. C. Schuldt, treasurer; along with nine other directors. Within three weeks 150 shareholders had purchased 330 shares, and on November 1, 1916, the Meriden Farmers Elevator and Mercantile Co. was formed. A mill to grind feed was installed and it was decided to sell coal, but patrons would have to unload it directly from the rail cars because no coal sheds were available. At the 1917 annual meeting it was announced that no dividends would be paid because expenses were too great. To improve the operation it was moved to purchase a larger engine to speed up handling of incoming grain.

Meriden Farmers Elevator, 1924.
L-R: the section car shed, tile in the foreground, salt and flour shed on the right.

Julius Bartz was named manager and remained until March 1921 when he resigned to go farming. There was no criticism of his management. The final audit showed that after losing money the first two years it had two profitable years but in the 1920-21 year it experienced a loss of $6,838.01. By November 1920, prices of all crops had dropped 48 percent from the previous year, and some grain prices dropped much more and were slow to rebound. The agricultural economy took the greatest tumble to that date in our history.

Paul Drache was enrolled in a two-year business course at Wartburg College when he was contacted about returning to Meriden to become manager of the elevator. The college agreed that if he took the position they would grant him his associate degree at the spring graduation. Drache became manager on April 1, 1921, at a salary of $125 per month. In his first year there was a loss of $1,878.01. I asked Dad how he managed to operate if the business continued losing money. He replied, "It was simple. I just sold more stock." The audit showed that during his second year there was a donated surplus of $3,216, which indicated that many of the stockholders were determined to keep the firm going. A slight gain of $36.17 was reported in the 1922-23 business year, followed by losses of $320 and $577 in the final two years of operation. Drache was hospitalized in Rochester in late 1924 and was unable to work full time but continued as manager on a reduced basis until 1926.

In 1925, an electric motor was installed, which speeded up grain handling. There was no record of anyone else being employed to manage after 1926, but the final audit shows that $3,250 was paid out for management. In the final two years $170,805.60 was paid for purchases, which yielded a profit of $8,163.78. The auditor's final report read, "The present condition of your company is due to the fact that the volume of business was not sufficient to produce enough profit to pay the necessary expenses of the business. Twice the volume of business could have been taken care of with the same amount of expense. The books and records were found in a very commendable manner and were very complete in every detail. In our opinion, the fact that at final closing, only four accounts amounting to $99.40 was charged off as a loss deserves commendation."

The words of Veblen expressed at the beginning of this chapter were proven in this case. Paul Buecksler, a life-time Meriden resident, understood the elevator's problem. He said, "Like most farmers in the area we never had grain to sell because we had cows, hogs, and chickens to feed. We had our grinding done there and bought concentrate to mix with the ground grain."

The July 30, 1926, *Journal* stated that Hayes Lucas purchased the elevator. It is likely that they considered doing so to avoid the competition in coal and tile, which were a significant part of their business.

In August 1926, Thomas Clark of Britton, South Dakota, leased the facility until May 1927 when he sold his lease to Arthur Willert who purchased it for $4,000 and operated it as Meriden Elevator. In August he purchased an attrition mill, which enabled him to do a better job of grinding feed. After the sale of the building, the Meriden Farmers Elevator Co. still owed $6,246.32, and the auditors recommended that the stockholders be assessed 100 percent of their original investment for the benefit of the creditors, which included the auditors' fee and a bank loan. It is my impression that all the creditors received their due.

Agriculture

Crane Creek/Drainage

Improving Crane Creek to secure better drainage and tiling were keys to better production for the farmers, but it also increased the tax base for the township, so everyone gained in the process. The plat map shows how extensive that system was in the township. To appreciate how important tiling was, Paul Buecksler was deferred four times by the Blue Earth

County draft board because of appeals from local farmers that he was needed to do tiling. He was finally drafted in April 1917. Upon returning from service in late 1918, Buecksler and his crew of fifteen were engaged to put in seven carloads of tile on the Carl Schroeder farm. Buecksler had a fifteen-man crew and he and his brother led the way by surveying where and how deep the tile should be placed. The next person dug a trench with an eighteen-inch spade. Then a second person, also with an eighteen-inch spade, followed and dug to the required depth. The spade men each received twelve cents a linear foot. The average spade man could dig about 250 feet a day, the very best up to 300 feet. Four men could dig 500 feet daily for six-inch tile "in good conditions." The main challenge for the diggers was to keep the spade handle clean so it would not interfere with their hands, hence, their rhythm. The standard charge for laying the tile in a three-foot trench was forty cents a rod. Tiling machines were in use at that time, but like most new technology, they are thought to be too expensive.

In April 1915, V. E. McDonald was contracted to dig a six-mile ditch to drain and clean out Crane Creek because of all the tiling that took place during the boom years. In 1918, a dredging crew started work on Ditch #5 on the Frank Kading farm. Tiling slowed down during the 1920s because of the poor farm economy, but by then everyone realized that the ditches had to be improved. In 1925, Harry Andrews, whose land was traversed by Crane Creek, called for petitions to have the creek re-cleaned because it was not handling water from heavy rains. The drainage program still had much work to be done, and Paul Buecksler became one of the prime movers in that process.

Dairying

The April 21, 1916, *Journal* noted that Ben Kuckenbecker was the only dairyman in the township to belong to the testing association, but others would soon follow. Dairying was making rapid progress in the county; it had twenty-four creameries in operation, which produced over 3.5 million pounds of butter that sold for $932,204.53. Dairying was the county's leading agricultural business.

In 1917, Henry Rosenau, the Meriden creamery operator, was recognized as one of the leaders in the industry in Minnesota but resigned after twenty-three years because of health problems. He purchased a farm in Woodville Township bordering Meriden and then acquired twenty high grade Holstein cows and a registered bull.

It is my opinion that Rosenau was one of the most progressive farmers in the Meriden area at that time. On a couple of occasions my dad allowed me to ride along with him and Rosenau when they hauled livestock to So. St. Paul. I enjoyed hearing him talk about farming because he was so positive. He was well read and I believe was a subscriber to *Doane's Agricultural Digest*, which he learned about while taking the short course for butter making at the university. The only other person my dad was aware of who subscribed to *Doane's* was Roy Bakehouse, a graduate of Iowa State, and I had the same experience riding with him. He also was so positive about agriculture.

Rosenau was a believer in diversification and in 1918 built a 100-foot-long hog barn. One of the unique features of that barn was that part of it was built in a T shape and at truck dock height to facilitate loading. Then, in later years, he also fattened cattle in that facility. He was an early customer of my dad's after he started trucking livestock in the 1930s, and Dad often commented that that building was one of the easiest and quickest places to deliver or load out livestock. Rosenau operated on a large enough scale so he always shipped a full load and often two loads at a time.

The high price for butterfat encouraged a rapid transition to cream separators. In February 1917, one weekly issue of the *Journal* reported that four farmers had purchased separators and named six other farmers whose cream checks grossed over $100. Fred Mundt had the highest check of $273.38. The size of the monthly checks kept increasing, topping at $503 for Henry Scholljegerdes in November 1920, just months before the price of farm commodities collapsed. (See the chart on the price of butterfat in the appendix.) The Scholljegerdes herd consisted of Shorthorns throughout his farming career, for he diversified his operation by fattening the bull calves. Shorthorns were a good dual purpose breed. His farm contained a large acreage of wood/pasture land, so he had ample land for summer grazing. He also purchased lightweight feeder cattle because he preferred feeding cattle rather than hogs.

During the long winter days, milking by hand with light supplied by lanterns was a twice a day, seven days a week task on dairy farms, Milking in the hot summer evenings with flies and having to avoid switching tails was not pleasant either. Initially, farmers purchased cream separators, which were less expensive and the skim milk could be fed to the calves or hogs, but they were reluctant to get milking machines because of the greater cost. Early ads stated that a farmer should have at least six cows to afford a milking machine.

As herds increased in size, keeping the larger volume of milk cool and having to deliver the larger volume was an added challenge. This encouraged farmers to buy cream separators and milking machines as soon as they could economically justify them. In the first week of March 1918, five township farmers purchased Hinnman cream separators and milking machines. The company boasted that sixty herds in the county were milked by their machines. By August 1918, the Meriden creamery had only twelve patrons who were still delivering whole milk and announced that effective in two weeks it would accept only cream because it was replacing its milk handling equipment.

Carl Enzenauer built a new barn in 1917 and completely modernized it with a power-operated separator and a milking machine, but he also installed all steel stalls and stanchions. In 1919, Emil Heinz, from the Crane Creek area, sold his complete herd of grade cows and purchased fourteen purebred Holsteins for $3,100. He joined Ed. Kuckenbecker and became the second farmer in the township to join the testing association.

In September 1921, the *Press* reported that "Steele County far outstripped all the counties in the dairy belt of Minnesota that are much larger [in area] in dairy receipts. [Some] of those counties have far more dairy cows." Steele County had 449 silos, 1,286 cream separators and seventy-two farms had electric plants. That year the county creameries had 2,015 patrons who milked 20,667 cows, which was sufficient to produce 3,227,493 pounds of butter for which the patrons received $1,474,448. It led all counties in butter production per square mile. "It produced more butter than the combined states of Alabama, Arkansas, Arizona, Connecticut, Delaware, Louisiana, Maine, Maryland, Rhode Island, South Carolina, and West Virginia." The Meriden Creamery was the fourth-ranking creamery in the county after Gilt Edge, Ellendale, and Moland.

William Scholljegerdes purchased a cream separator, and a double unit milking machine in 1922. The separator was operated with a one and one-half horse-power gasoline engine, and the milker required a two and one-half horse-power engine. "They were both real time savers but electricity, which we got in 1928, was even better." Scholljegerdes commented that he was a loner and both he and his wife Marie preferred not to have hired help around, so they automated as fast as they could.

In 1924, L. E. Radke, who was one of the leaders in dairy and hog production, installed water bowls for his herd. He was probably the first in the township to do so. The lack of adequate water was a major reason for low milk production, but in Minnesota winters it took good buildings to prevent the water from freezing, so water cups were slow to be adopted.

By 1924, more farmers were testing their herds' production, and the paper started publishing the monthly scores. The Nielsen-Winship farm, near the eastern edge of the township, had a herd that averaged 701 pounds of milk and twenty-seven pounds of fat for the month of November. The top cow produced 1,176 pounds of milk and forty-one pounds of fat. At that time the average annual production per cow in the United States was 3,527 pounds, compared to Canada with 3,779 pounds. The Netherlands was the leader with 7,585 pounds. William Hammann joined the testing association and culled his lowest eight producing cows, leaving him with twenty-six cows, but that was still the largest herd in the county association. His highest producing cow led the county with 1,395 pounds in April 1925 and fifty-eight and one half pounds of fat.

Roy Bakehouse had a degree in agriculture from Iowa State and was employed to manage the National Farmers Bank farm south of Owatonna. That herd had the highest average in the No. 2 testing association for grade Holsteins with fourteen cows out of 358 in the association with 1,069 pounds of milk and thirty-seven pounds of fat. His highest producing cow had 1,626 pounds of milk and seventy-four and eight tenths pounds of fat. He had competition from H. J. Bundy, from the Crane Creek area and also a graduate of Iowa State, who managed the Ayrshire herd on the Nielsen Winship farm.

Harvey Kuckenbecker, son of Edward, and Clarence Bundy were both active in showing prize heifers from their fathers' herds and were regular winners at the county fair, the State Fair, and also at the Waterloo Dairy Cattle Congress. In 1925, Kuckenbecker was the only Minnesota entry at the National Dairy show in Indianapolis where he received the grand champion award in the Holstein class. At that time he was in the eighth grade in District 36. That spring he earned $50 in cash, a gold medal, a $1,000 Minneapolis *Journal* award to the Agricultural School in St. Paul, plus several other cash prizes. Clarence Bundy took similar honors in the Ayrshire class.

In 1927, the Steele County Holstein Breeders Association had 132 members, the largest in the state association, which had a total of 400 members. The Bakehouse, Bundy, Heinz, Kuckenbecker, and Radke herds were the leaders. They regularly produced the top figures in dairy production and in showing others what could be done, but it was not until the 1940s that more Meriden farmers became involved with testing.

Two other events took place at that time that enhanced the dairy industry. In 1926, the Meriden creamery purchased a buttermilk drier, which improved its income from a by product that was sold on bids to the farmer who was willing and equipped to haul it to his farm and use it for hog feed. The food processors had learned how to incorporate dried buttermilk into other foods, and the creamery received four cents a pound for the dried product. In one month it grossed nearly as much as the wet product paid in a year. The second change that took place was that state veterinarians entered the township and tested dairy cattle for tuberculosis. In December 1926 the Steve Wodarczak, Ernest Drache, and H. J. Bundy herds were accredited free of the disease.

In May 1928, an announcement was made that several veterinarians would be brought into the county so all the cattle could be tested in a fifteen-day period. This was the first step in putting the county on a modified accredited list toward becoming a recognized bovine tuberculosis-free county. Out of the 1,809 herds and 29,576 cattle tested, 1,303 cattle in 494 herds were tubercular. William Scholljegerdes lost ten head in that test but quickly rebuilt his herd from his own stock. The diseased herds were retested in November and March, and all the herds would be retested in July 1929. In August 1928, four carloads of cattle that were rejected were shipped by the Meriden yards.

Record Keepers

It is well known that farmers traditionally kept their records in a cigar box or a shoe box and at the end of the year added the expenses and the income to determine if they lost or made money. Some felt that if their bank account had more money in it at the end of the year than they had at the beginning that they had made money. Because of the dearth of data, the extension service was at a loss as to how to help farmers improve their management practices. Fortunately, the Morrill Act provided for the establishment of experimental farms that started test plots. Record keeping was also part of the program. In 1894, Willet Hayes, the director of the Minnesota experiment station, wrote, "In the progress of agriculture, system and good management are not as highly developed in agriculture as in our other great industries." To be of help to the industry the extension personnel had to know more about how farmers actually derived their income. The organized research in farm management was started at the University of Minnesota, and the state became one of the leaders in that discipline. In 1915, Dr. George Pond started research in farm management and is recognized as the father of that discipline in Minnesota. Farm institutes were started by the extension service, and the railroads became involved because they realized that agriculture was becoming commercialized and that would greatly help the fledgling food processing industry. Extension personnel and county agents worked to establish groups of farmers who were willing to keep records of their farming practices.

The first accounts of any record keeping activity in the county were published April 15, 1921, when data from 22 record keeping farmers was revealed. That year the county had 1,860 farmers, 1,413 of whom were owner operators and 447 of whom were tenant operators. Seventeen of the 1,860 had already become tractor powered, and the others had an average of 5.69 horses or mules and an average of 14 dairy cattle, which means an average milking herd of about 7, and about 16 hogs per farm. The 1,860 farmers raised about 14 acres of corn, which averaged 42 bushels an acre. This was 15 bushels greater than the national average. The second largest crop was wheat, and the average farm in the county had 10 acres, which yielded about 10 bushels per acre. The average receipts were $1,727 for dairy, $795 for swine, $184 for crops, $148 for poultry, and $18 for outside work for a total income of $2,947, less expenses of $1,613, leaving $1,334 cash to pay capital cost and for family living. These figures indicate that the majority of the farmers were not doing well, but 1921 was a historic low year in the farm economy after nearly two decades of boom years.

In 1922, the 23 farmers in the record keeping group had farms that ranged from 80 to 336 acres, for an average of 182 acres with 132 in crops. The average work day of the farmer was over 11 hours, and that of the unpaid family labor ranged from 15 minutes to as high as 30 hours, proof of the adage that many farms survived by the unpaid labor of women and children. Of the farms that employed labor, the average was 132 days annually. The gross receipts of this group of 23 record keepers averaged $4,049, less expenses of $2,195,leaving a net of $1,854, which included 2,499 hours of family labor valued at $437.

In 1924, the 23 record keeping farms averaged 190 acres, with an average of 41 acres in corn, 31 in small grain, 27 in hay, and 49 in pasture. Those farms had a gross income of $5,507 with a net return for labor, management, and capital of $2,463. Dairy, crops, hogs, cattle, and poultry, in order, were the five sources of income. An article on June 5, 1925, gave the results of the record keeping farmers and "indicated a need for larger farms. Their average for 1924 was, $4,471, which is larger than the average for the county and most of Minnesota. They had an average expense of $2,246, leaving net cash of $2,025."

In 1923, the Meriden Farm Bureau, with aid from extension service personnel, made a strong effort to establish dairy and poultry enterprises along with increasing alfalfa acres and securing ten purebred dairy sires in the township. One of the monthly programs featured a three-reel motion picture on poultry raising, marketing, and egg production. As a result of the poultry programs, Albert Nielsen purchased a 600-egg incubator and Louis Radke a 1,200-colony brooder so they could increase their flocks. The *Journal* told about Adolph Schmidt, of the township, who started with

200 acres in 1903 and soon had "19 milk cows housed in a large barn with all steel equipment, ventilation, and a water system." He farrowed 200 pigs from 25 sows, and Mrs. Schmidt had a flock of 1,400 Barred Plymouth Rocks and White Leghorns and had as many as 70 setting hens at the same time.

In Owatonna, the Rotary Club heard a message from F. E. Young, principal of the OHS, entitled "Over Population May come in Half a Century." Young pointed out that it took two and one-half acres per capita to feed the nation, more than any other nation. The nation had been doubling its population every forty years with only 50 percent of the tillable acres being cultivated.

But farmers were paying attention and started improving their management. For example, in 1919 there were only 93 acres of alfalfa in the county, and by 1926 that figure had increased to 1,350 acres. As soon as farmers started increasing their alfalfa they became more aware of a need for lime. In 1924, the farmers pooled their orders to buy a carload of lime from Mankato. L. Radke, A. Schmidt, L. Stendel, and F. Nass, whose farms were all adjacent, purchased a carload of lime. During those years agriculture had gone through a serious financial shock as shown by the following data from the *Journal Chronicle*

TABLE I

Receipts	Farm Income	Interest	Return to labor & mgt.
1920	$5,492	$597	$1,723	$1,126 loss
1921	$3,509	$28 loss	$1,656	$1,684 loss
1925	$5,963	$2,940	$1,091	$1,894 profit
1926	$6,880	$6,306	NA	$3,317 profit

In 1925, The Master Farmer movement was organized as a means of dignifying agriculture. In January 1929, about forty farmers who belonged to the Better Farming Club in the five-county area said that they profited by being a member of the group and would continue to belong. The Steele County extension agent and the extension field man for that area tried to get twenty new members.

Specialty Crops

In October 1916, thirty-three cars (about 400 tons) of wire grass, were shipped from Meriden to the Willow Grass Co. of Green Bay, Wisconsin. It had to be harvested with a scythe or an older model reaper and then bound by hand so it would not be bent. This was necessary so it could be made into rugs, mats, carpets, and other articles that use vegetable fiber. That fall Max Drache and Herb Roseau sold their flax straw to Andrew Erdman, who baled it with his stationary baler. Two cars were loaded at Meriden, along with flax straw from others in the township, to be shipped to a textile processor. Both of those products were in demand because of the shortage of other textiles due to the war in Europe. There is no record of the willow grass being harvested in future years, and only a few farmers raised flax after they expanded their livestock production.

By 1916, the Owatonna Canning Company was large enough that it was able to wholesale its own corn, peas, pumpkin, and tomato production. In 1926, the company contracted for 700 acres for sweet corn and 400 acres for peas and small acreages of other crops including pumpkin and squash. That year L. C. Lange and Henry Hartle, owners of the company, purchased the Sommers Cannery, of Kenyon, and operated it in conjunction with the Owatonna facility. The company canned 40,000 cases of peas out of a total of 485,139 cases of peas from all eight canneries in the state. In 1928, the cannery had six viners stationed in the growing area. The two nearest to Meriden growers were located on the A. P. Bartsch and John Ebeling farms, both near the east township line. Over 55,000 cases of peas were processed by two shifts, each with eighty people.

Threshing crew L-R Martin Altenburg, Bill Benhsman, Louis Scholljegerdes, Herman Frahman, Louis Palas, John Krippner, and Henry Palas. Palas owned the thresher and engine. This crew all worked on the rig when it operated. C.a. 1922-1925 Virena Palas Photo.

The June 19, 1925, *Journal Chronicle* reported that sugar beets were produced on the Charles Jones farm near Owatonna. Jones had grown beets the previous year and had harvested eleven and six-tenths tons of beets and a two-ton crop of tops for stock feed. In 1926, W. R. Whelan, H. W. Schuldt, Herman Brase, Carl Enzenauer, and Mike Gontarek raised and shipped thirty-three carloads of beets to the American Beet Sugar Company at Mason City. Beets are a very labor intensive crop, and Mexican families were employed to do the hand hoeing, topping, and harvesting. The crop was sufficiently successful that the company signed up more growers. In 1928, 200 acres of beets were produced by Herman Schuldt, 25 acres; H. H. Grandprey, 10; Herman Brase, 20; William Whelan, 25; L. P. Zimmerman, 20; Bill Beshman, 35; Edward Ruedy, 35; and Nick Ohnstad, 30. That year the company installed a stationary dump in Meriden just south of the tracks across from the depot to facilitate shipping of the expanded production.

More Mexican workers were needed. During the summer the men often came to Harvey Moore's pool hall. I remember the excitement in the village when fights broke out and Moore, with the help of other local residents, had to intervene and stop a game. Moore told me that there was even some minor stabbing during pool games. The most serious incident that summer occurred on the O. H. Schueller farm north of the village when two of the workers attacked a third person, "slashing with knives, a razor, and a hatchet. Hearing was delayed pending Castello's recovery."

On October 1, 1929, beets started to arrive at the dumping station, and by November 10 a total of sixty-four carloads had been shipped. The beet workers went to St. Paul and Kansas City, except for Berti and Hernandez, who were found guilty of second degree assault and were sentenced to one and one-half years in Stillwater.

In 1926, the first commercial hybrid seed corn was developed by Henry Wallace, which is still recognized as one of the singular most significant breakthroughs in world food production. In 1927, Minnesota 13 and Golden Glow seed corn was available in the county. Farmers were advised to study their corn in early September before it was husked. "The ears should be somewhat long and small in circumference and heavy for their size. Ears with sixteen to twenty rows of kernels are preferred and there should be no discoloration. The corn stand of three eight-ounce or foursix-ounce ears per hill at three feet six inches should yield seventy-five bushels per acre." This was nearly triple the national average of twenty-six bushels, which implied that the extension people realized that Steele County had great potential for corn production.

A full page ad in the March 23, 1928, *Journal* stated that smut in small grain crops in 1926 had infested 70,234 acres, resulting in a loss of $100,000. Farmers were informed that they could treat their seed with copper carbonate dust or a formaldehyde treatment prior to planting. This was probably the first mention of any chemical aid to farmers for any crop in the area other than arsenic for killing potato bugs. At the Corn and Field Day on August 14, 1929, sponsored by Minnesota Crop Improvement, modern corn breeding was explained and observations were made of the experimental plots. In 1930, the county received some double-crossed corn seed for test plots.

Threshing

In August 1915, the township farmers were battling heavy rains in their effort to complete harvest. The *Journal* reported, "Because of heavy rains gasoline engines are being mounted on the binders to furnish power to run the sickle [because the bull wheel was sliding instead of turning] so all the horses had to do was pull the binder. The demand [for engines] is so great that manufacturers cannot meet the demand. The implement dealers have been stripped of the old fashioned cradles for cutting the grain because they were low on inventory because they had not replaced them from their previous wet year." The dealers probably never thought anyone would ever want a cradle scythe.

In 1915, the township had at least three threshing rigs, but it is not clear whether the threshing was done from stacks in the yard or from shocks in the field. In June, Gust Fette and Herman Schuldt purchased a "threshing outfit," and a group of neighbors organized a "threshing run," which meant they cooperated to haul the shocks of grain from the field and helped to put the threshed grain in storage. The rig went from farm, to farm and the farmer where the threshing was being done had to provide meals for the entire crew.

My grandfather, Max Drache, became involved with a "threshing run" between 1912 and 1915. This "run" was loosely referred to as the South Meriden threshing run in which the threshing machine and the tractor were owned by Theodore Pump, and the grain elevator and two grain teams were owned by Max Drache. The threshing machine was a forty-two inch wooden Nichols and Shepherd, and the tractor was a 35/70 Minneapolis Moline. The wheels were close to eight feet in height, and it had a roof over the operator area. The elevator was a wooden Sandwich flight model that reached to the peak of the granary from which the grain was then diverted to any of the bins within.

The original members of that "run" were Alvin Abbe, Herman Abbe, John Abbe, Martin Altenburg, John Buelow, Max Drache, August Gross, Orlando Kading, Leo Lebahn, Andrew Matthes, Theodore Pump, and Albert Stendel, all of whose farms were virtually adjacent.

Each year in the meeting prior to threshing the big issue was determining the order of threshing. Drache and Pump kept records of the order when each farmer threshed every year and made sure that everyone was in proper sequence. This was important, especially if there was any rain during the days of threshing. After threshing a settlement meeting was held in which the finances were equalized. Every farmer had to furnish a man and a bundle team. The larger farmers had to pay the smaller farmers because the smaller farmers had to work more hours at the larger farm places than at the smaller places.

The threshing crew consisted of ten bundle teams and two to four spike pitchers in the field. Pump oversaw the threshing machine, Herb Schroeder was the engineer, and Drache operated the grain elevator. It took two men to handle the grain wagons and one man to work in the straw pile. Sometimes a person tended the blower, but generally the crew was made up of sixteen to twenty men. Some crews were as large as thirty.

My father recalled that in the days before the elevator his job was to fill the two-bushel Bemis sacks in one grain wagon while two men hauled the full sacks and hand dumped them in the granary. At other places the grain was threshed directly into the grain wagons. Then, the men shoveled the grain through windows in the side of the granary.

At most farms three women were required to prepare the meals. My aunt, who helped in the house in the late 1920s up to 1934, recalled up to twenty-two at the table. The crew was fed morning lunch, dinner at noon, and afternoon lunch. Threshing was an important event every year and much of the conversation was about the "good or not so good meals" that were served at so and so's place.

The Meriden correspondent wrote in her August 1916 column that several threshing rigs were operating in the area, and more shock threshing was being done than ever before. "The scarcity of laborers and an ample supply of threshing machines have been two causes that have brought about the change." After combines came into existence the number of participants dwindled until the labor intensive threshing bees came to an end.

Silo filling was another event that required the cooperation of neighbors but that did not demand as large a crew because of the silo filler's limit to handle more volume.

Tractors and Implements

In 1892, the first successful gasoline tractor was invented, and by the 1920s tractors proved to be far more efficient than horses. But the horse people fought a long, losing battle against tractors. They developed a twenty-horse hitch to pull a modified eight-bottom tractor plow that was capable of plowing eighteen to twenty-two acres a day. To care for those horses required 2,800 hours of labor annually, and they consumed crop from about 100 acres. Except for a horse raiser like John Tuthill, no farm in the township had that many horses. On May 5, 1916, the paper advertised that tractor demonstrations were being held in the county.

The first commercial potato diggers appeared shortly after the Civil War. Like potatoes, wheat accompanied most of those who went to the frontier because it was so adaptable. In 1917, John Steinberg purchased a potato digger, which he used to help his neighbors. Nearly every farm in the township had at least an acre or two for personal use, and many planted an extra acre or more for income.

In 1917, Herman Schuldt purchased a tractor, which may have been the first one in the township. Henry Ford adapted his successful assembly line to his tractor factory and effective January 27, 1922, the Fordson price was reduced from $625 to $395 FOB factory. Henry Ford was quoted, "The horse on the farm is a living expense, he eats more than his worth, [and] he must be done away with."

The Fordson was popular because it was smaller than most tractors and was what farmers wanted—something that would pull a two-bottom plow and the implements that matched it. Prior to 1922, most tractors sold for $1,000 to $3,000, but Ford was desperate for funds so he adapted the assembly line to cut the cost of production. When he reduced his price from $625 to $395, orders ballooned. By 1925, he had sold 104,168 tractors, which was one-fifth of all tractors that existed in the nation.

Henry Palas purchased a Fordson when he started farming in 1925. His wife, Virena, recalled, "That was about $400 and when that wore out we just got another one." The Palas farm was just east of the District 48 schoolhouse, and I remember the whining noise the tractor made as we watched it travel back and forth across the field. On one occasion I saw the front end rise up when the plow must have hit a rock. In addition to the Fordson, Palas also had two horses. He never farmed more than seventy-two acres but they maintained a milking herd of twelve.

In 1925, August Beshman purchased a tractor, and in 1926, Henry Scholljegerdes purchased a Fordson, as did Carl Brase. Anthony "Tony" Gontarek purchased a new "light duty truck" and worked for a contractor hauling gravel.

In 1920, horse numbers for the nation peaked at about 27 million. That number declined sharply to 16 million by 1925 because farmers quickly saw the advantages of the gasoline engine. But in February 1928, the ex-

Henry Scholljegerdes operating his new Fordson with a single row mounted corn picker in 1926, which was reputed to be the first of its kind in Steele County. Robert Scholljegerdes photo.

The 1926 corn crop in storage in a woven wire fence. The corn was ground and fed to beef cattle during the winter months. Robert Scholljegerdes photo.

tension people pointed out that despite the competition with tractors, there was still a strong demand for good quality horses. They stated that there was a shortage of draft horses in the eastern Corn Belt. "A sensibly limited breeding project might be worthwhile. The number of draft horses had declined since 1920 and farm numbers fell right with that."

In 1929, a study by Dr. George Pond on the scope, cost, and potential of 100 Steele County farms indicated that in 1927 there was one tractor for every five farms in the county, and one tractor for every 330 acres. Throughout 1927 and 1928, the Owatonna Ford car and tractor dealership met with virtually every Farm Bureau unit and showed two films about how tractors were built on the assembly line and sold for $495 FOB, and how the new Model A cars were built. Those films were followed by a comedy film. After 1928, when that tractor was discontinued, nearly 850,000 Fordsons had been sold—"the most popular tractor model ever built in the world [up to that date]." In 1924, International Harvester introduced the Farmall all-purpose tractor, which could do row crop operations and would end the horse age of agriculture.

Leo Lebahn with his 1925 Model T pickup and trailer with nine crates of their first crop of chickens being delivered (1927) to the Kelly Company in Owatonna.

In September 1926, Leo Lebahn and Mollie Drache, my dad's sister, were married. They were financed by my grandfather to start farming. They purchased four horses, $420; cows, $675; hogs, $175; chickens, $50;

harness, $145; two wagons, $140; cream separator, $120; manure spreader, $150; gang plow, $100; corn planter, $175; grain seeder, $175; hay, $100; seed oats, $95; succotash, $50; total $2,570. In 1929, my grandfather helped his son, Otto, get a start on a 120-acre farm in Havana Township. This included a used car, $250; machinery, $1,340; oats, $40; succotash, $15; cash, $2,600; total $4,245. Neither of these two couples ever expanded their operations.

Electric Power and Light

In January 1915, Sam Grandprey added an 8 x 12 foot washroom to his house and installed a gasoline engine to run their 1898 Maytag wash machine and wringer. "Mrs. Grandprey said it worked to perfection and is a great help." In March 1916, Grandprey installed acetylene light systems in his house, the Grandprey Hall, the blacksmith shop, and the post office. Apparently this was the first non-kerosene lighting in the village. In July 1918, Joseph Silkey sold his farm and purchased an electric plant in Claremont. That year Theodore Pump, John Brase, and Henry Rosenau had Acetylene Underground Generator systems installed for lights, cooking, and ironing. Rosenau extended his system to his farm buildings. Rosenau was still using that system as late as 1937 when he lit the burners in the beef barn prior to loading cattle. The Rosenau farm was off the main graveled road and installing electricity was too costly. In 1919, St. Paul's Church installed an acetylene system, which was probably one of the last of those sold in the township.

By 1918, Delco Light plants were advertised as being capable of running fanning mills, grindstones, cream separators, butter churns, washing machines, and "other light machinery." The Delco system relied on storage batteries that which were charged by wind towers. In 1920, Henry Luhman put those lights in his store, in the livery barn, and one in his yard. L. C. Palas installed that system in his garage just north of the tracks and the large house—the second one north of the garage. Fred Fette did the same for his large house and combination barn and garage on Mill Street.

In 1923, after years of waiting, electric power was installed in the village by Howard Deichen, the owner of the power company in Waseca. Paul Buecksler and his brother were employed by Deichen to survey for an electric power line east from Waseca along the road north of section seventeen up to the east township border. When that was finished they surveyed south of Meriden to the first road running east. My Aunt Adeline recalled that my grandfather paid $500 to be connected. They had a ceiling light in every room. The farmers who lived along the two lines that Deichen built received their power in 1923, and the next year farmers had meetings to request service.

My aunt remembered that when the crew installed electricity in the basement they helped themselves to the home brew. During the heat of summer, water from the windmill pump flowed into the water tank in the separator building. From there it drained into the cattle tank so water in the tank in the separator building was always cold, and home brew and home-made root beer were always there. The cement block separator room was always cool so it was a very refreshing experience to reach into the cool water and have a bottle of root beer while the adults had home brew.

By November 1926, Northern States Power had extended its line west of the Golden Rule Creamery into Deerfield, and in April 1927, the Owatonna Municipal Electric Utility agreed to run lines into the county where farmers were organized and wanted the service. Within a month the Interstate Power Co., from Albert Lea, agreed to construct a line to Medford and to develop farm lines from that village. It also agreed to run lines to other villages, but it was not allowed to enter Owatonna. Lemond and Berlin townships were served by Interstate, but because they were not on a main road it was not until 1928 that the Henry and William Scholljegerdes farms were connected. William, who had his water system, separator, and milk machine powered by gasoline engines, commented, "We had to pay $500 to hook up and $800 for all the wiring and motors, but electric was easier to work with and a real time saver. We never objected to the electric bill."

That year Edison Electric Company, in conjunction with Deichen of Waseca, sponsored a cooking demonstration to thirty women about using electric stoves. Miss Fern Luhman, Mrs. Rudolph Schendel, Mrs. O. P. Jensen, Mrs. Art Willert, and Mrs. Paul Drache were all winners of prizes for their preparations. All the prepared dishes were served at a lunch that followed the program. In 1929, thirteen farmers in Deerfield organized and signed up for service on a line that extended from the Bert Buscho place to the O. E. Wilkowski farm; both were patrons in the village.

Farm Organizations

Abraham Lincoln made the statement that politicians catered to farmers because there were so many of them. Up to the late 1800s, farmers made up nearly half of the labor force. Farmers had prided themselves as being rugged individualists and were inclined to avoid getting involved with organizations, but as their numbers declined so much they felt threatened because they were no longer in the majority. This was especially true when the golden years came to an end after World War I. In 1915, the Nonpartisan League was organized in North Dakota and became the most radical political agricultural movement in the Midwest. By 1919, they had entered southern Minnesota and "many from Meriden" attended a League-sponsored picnic at St. Olaf Lake. A former mayor from Minneapolis was the featured speaker, and the Meriden Brass band provided the music. In August 1920, about fifty attended the League meeting at the town hall where Miss Lillian Anderson, from Minneapolis, spoke. In June 1922, about 500 attended a picnic sponsored by the Meriden Women's Club at the Gottlieb Haas farm. "Lillian Anderson gave a very interesting talk on political affairs." That was the final news from the paper about the League.

In May 1921, the county agent and three township farmers conducted a two-session meeting at the town hall. The topics were quite diverse. The afternoon session explained what the Farm Bureau stood for, and the evening session was about how farmers could benefit by having silos. In October 1921, township Farm Bureau units were organized in Clinton Falls, Deerfield, Havana, Meriden, and Merton. The Meriden director was William Whalen; vice director, Albert Nielsen; and secretary, Maxwell Andrews. At that time the county Farm Bureau already had 1,030 members. A debate at one of the unit meetings in the township was "Resolved that a Farm Raised Couple will find a more Desirable Lifework in Farming than in a City Occupation."

In 1922, the Bureau was instrumental in the formation of the Central Cooperative Oil Association, which was organized by 160 county farmers. William Whalen, of the township, was a director, and A. P. Bartsch, the president. In June 1922, about 3,000 attended the Bureau picnic in Dartt's Park. Several speakers were heard after which there were various sporting events. The oil company grew rapidly. By 1926 it had 586 members and in 1929 that figure reached 1,200. It was stated earlier that Standard Oil had withdrawn its bulk station from the village, probably because of the strong competition from Central Cooperative.

In 1928, a speaker at the Better Farm Course said this that was a good time to buy land but buyers should be careful not to get too indebted. "We are in a highway and school building era, we have a problem because too much public revenue is levied on property taxes and the amount farmers pay is inordinately high." The big news of 1929 came from A. P. Bartsch, one of the most prominent farmers in the western part of the county. "He decried the trucking trend which is a menace, and directly threatening the [shipping] associations' welfare."

Daily Life and Culture

Automobiles and Roads

When William Scholljegerdes and Marie Krenke were married in 1911, Ewald Lewer, the Dodge dealer from Waseca who grew up on a farm in the township, gave them their first automobile ride. Their daughter, Gertrude, said "that gave them car fever." They lived just across the line in Lemond Township but were very active in Meriden, so their automobile was one of the earlier ones in the Meriden trade area. In 1915, the Ford Motor Co. presented a film that featured the universal car, the Model T, which was made on an assembly line that was created by Henry Ford in 1914. The film showed how 16,000 men at the plant working in crews of seventy on a 300-foot assembly line could produce a car every thirty-eight and two-fifths minutes and produced 1,000 cars per day. Ford made this film available throughout the nation. It was presented at the Metropolitan Opera House in Owatonna.

By 1915, Minnesota had 86,000 licensed cars. Steele County had 1,003 of which 599 were in Owatonna. The first notice in the paper of a car owner in the village appeared September 24, 1915, when H. J. Luhman made an automobile trip to Minneapolis. The first fatality in the township caused by an automobile happened along Highway 14 in November 1915 when Leona Beers was struck. Three children were playing on the road, and Leona apparently jumped in front of a car. Another accident in the township took place when the tongue broke on the wagon driven by Peter Eliason and the horses became frightened. The driver of a Ford car that was following became frightened and drove his car off the road and into a fence. That year farmers in the county had about 500 cars.

The township was still using the poll system in lieu of cash, but in 1915, $3,000 was levied for a new building for the road equipment, plus another $800 for dragging the existing roads. In December, eight to twelve farmers with teams hauled three to four loads each day from the Anderson gravel pit "and would continue as long as the weather is good."

In February 1916, Theodore Pump, who owned a tractor and a road grader and did road work for the township, reported that the township had done much grading in the previous three years. It had sixty miles of roads that were not on ground good for road building, and there was no gravel available.

W. A. Schultz operating W. R. Schultz's Austin Mammoth Senior Grader forming a road in Lemon
Township, which became county road No. 7.

Teams were still allowed on major roads as was proven by Martin Altenburg, accompanied by Theodore Pump, who drove a team of mules and wagon from Meriden to Balsam Lake, Wisconsin, a distance of 150 miles. Sturdy mules and the cool November weather enabled them to deliver the rig to Altenburg's son-in-law in two and one-half long days. By comparison, wagon trains acrossing the virgin prairie traveled about fifteen miles a day.

The village had no gravel on its streets until 1916. That year Albert Reiter had a Reeves 22 tractor that he used to pull the township grader, the machine that was used to build six miles of new road. He boasted, "Meriden is going to keep its roads in as good condition as any in the county." That grader lasted until 1926 when the township purchased a new, improved grader. Frank Domy's son, Harvey, left the farm that year to attend an automobile school in Detroit. The final report for 1916 appeared in the December 29 paper simply stating that the roads into Meriden were choked with snow drifts five feet high and twenty feet across. In 1921, Minnesota adopted the Babcock road plan, which called for a system of taxation. In 1923, Steele County voted to allot money to each township for roads.

Due to a rise in crude oil prices, the cost of gas rose to sixteen and eight-tenths cents a gallon for low test and nineteen and six-tenths cents for high test. Regardless of the increased price of gasoline, people insisted on driving. In March 1926, the state allocated $1,066,692 to the county for that period but only $356,655 had actually been collected in the county. In 1926, the state designated two state aid roads in the township, "one from Highway No. 7 south to the church then east to section twenty-five where it connected to the Lemond road which runs at an angle out of Owatonna. The second road ran north from Highway No. 7 to state aid road twelve at a point that runs into Medford." In August, the graveling of route No.7 was completed as was the bridge over Crane Creek; all signs were in place. The total cost for graveling was $20,577. The Railroad and Warehouse Commission then placed stop signs at all three railroad crossings in the township "where motorists are expected to stop before passing over the tracks."

In 1921, most of the cars were put up on blocks for the winter, but by 1926 Minnesota had 624,527 cars, and 90 percent of them were used year round. The highway department could not keep up with snow removal demands and complained the "gravel roads were just not capable of the excess usage." Gertrude Scholljegerdes Voss said her mother learned to drive but had a "slight fender bender while in Meriden and never drove again." Her parents were anxious for her to learn to drive so she could make the four-mile trip to Meriden with eggs and buy groceries. In January 1927, twenty men with shovels and some horse plows opened the snow-clogged road from the village to the highway. That was a forerunner for what was to happen in April when all roads, except the paved highway to the north out of Owatonna, were impassible because of snow and soft conditions.

The big news locally for 1927 was the announcement that electric poles south out of the village had to be moved so a new, wider road could be constructed. The second big news was that the state wanted to extend Highway No.7 across "the richest section of the state," all 290 miles from the Mississippi to the South Dakota border. "Rochester, Owatonna, Mankato, and New Ulm are all shaking loose from the mud." As part of that project the state built a new garage along the highway just north of the CNWR tracks on the west side of Owatonna. That brick building, which still stands, was built at a cost of $19,000.

Two New Rural Schools

In March 1916, the village school, under the direction of Eva Ochs, sponsored a basket auction. The admission was fifteen cents, but the bidding was good and $84 was raised. The same event in the same week at District 52 only produced $15. Miss Ochs was not so lucky in February 1918 when she was quarantined with scarlet fever and District 48 was closed for two months. That year the village school had fifty-three students and, in keeping with the war effort, they were all enrolled in the Junior Red Cross. In the fall all the schools in the township were closed because so many people were ill with influenza.

When school opened in the fall, District 48 had fifty-eight students, so in May 1919 the residents voted twenty-five to four to give the board the authority to select a site for a new two-room school. Two sites were chosen—one across the street from the creamery and the other just south of the current school. Nearly as many women as men voted thirty-seven to twenty-one for the north site. The state architect and the rural school commissioner visited and approved the site but required that two acres were necessary for a two-room building. No bids were received to contract for building the school, so the state granted permission to rent the Evangelical Church building. They would receive state aid, but the board decided to improve the former post office building for the primary grades and the upper grades would remain in the schoolhouse. Dorothy Warren was the primary teacher and James Harty taught the upper grades.

In September 1919, a contractor agreed to construct a 60 x 64 foot building with two classrooms and two basement play rooms for $7,000. The building was not completed for opening in September 1920, and by then the cost was set at $16,000 for a modern brick building with folding doors that opened for a large auditorium with a modern lighting and water system. School opened November 1, 1920, but the school grounds were still torn up. There was no information as to the delay or the cost overrun, but those in charge must not have been upset, as only $2,000 was levied for school purposes—$1,000 for operation and $1,000 to apply to the building. The year was finished, with the two-day rural school state examinations for spelling, grammar, composition, elementary American history, arithmetic, elementary citizenship, and geography. Even after the new school was finished the traveling county library continued to send books from the Owatonna library for the school and also for the public, who received their books at the Hayes Lucas where Emil Buboltz, the manager, acted as the librarian.

Construction went much better for District 21, where the contractor started in April and on September 14, 1928, the dedication ceremonies were held. Werner Brothers from Austin were the general contractors. They roomed and boarded at the Paul Drache home (my parents). A public program was opened with invocation by Rev. Harrer. William Gibson, editor of the *Waseca Journal*, was the featured speaker followed by musical numbers. A luncheon was served in the basement.

Beyond Rural School

In the late 1890s, a few from the township, especially those living on the eastern sections, started attending OHS, the Normal School, Canfield

Business School, or Pillsbury Academy. After the turn of the century, Harry
Andrews, Lloyd, Medora and Harry Grandprey were some of the first to
seek education beyond the eighth grade. Others were Anna Schultz, my
mother, who grew up in Lemond Township, took her first seven years of
school in District 54, and then spent a year in St. Paul's parochial school,
after which she went to Lincoln School in Owatonna to finish her grade
school. In 1917, she entered Wartburg College and enrolled in the two-year
commercial course. In the spring of 1918, she took a job at the Security
State Bank of Owatonna for $42 per month with the intention of returning
to school in the fall. The bank had lost several employees to the military,
so when fall arrived she was persuaded to stay. Her salary was increased to
$85 per month. She always regretted that decision because she wanted to
be a parochial school teacher or do mission work.

Paul Drache, my father, graduated from District 43 and knew that he
did not want to become a farmer. His parents realized that too, so in 1915
he went to Canfield Business School. In 1916, he got a job at the First State
Bank in Meriden where he received $10 per month. He said getting the job
at the bank was a no brainer because his aunt was married to the president of
the bank board. Lloyd Grandprey also worked at the bank as assistant cashier. Drache did most of the work on the stockyard account, which included
writing out the checks. He had taken the handwriting course at Canfield
and was a good penman. In January 1917, his pay was increased to $15
per month. Later that year ,when Grandprey was drafted, he had to assume
some of Grandprey's responsibilities and his salary was increased to $25.
In the fall of 1919, he enrolled in the two-year commercial course at Wartburg College but, as mentioned previously, before he finished his course he
was called to manage the elevator.

In the spring of 1924, five farm boys from the Crane Creek area enrolled in the agricultural short course at OHS. That fall eight other students
from the Bundy, Nass, Radke, and Schmidt families enrolled in OHS. They
car pooled during the good weather and stayed in town during the worst
weather months. Clarence Bundy was one of those, and before his high
school career ended he won "highest national honors as a judge of hogs."
He was also on the team that took fourth place in junior livestock judging
at the International Livestock Exposition in Chicago. Later, Bundy was
named grand champion boy club member, plus he exhibited a grand champion Aryshire of its class. Four of that group of eight continued on for a
college education.

In 1925, only twenty-five students graduated from the county rural
schools. Of that number six were from Meriden Township schools, a low

point for rural graduates. The following year forty-seven graduated from the rural schools, "which was remarkable considering that the passing grade in the state exams was raised from 65 to 75 percent on the four major and two minor subjects." The county superintendent had set a goal of 100 graduates. In 1927, the rural schools had their largest class in the history of the county when eighty-five graduated at the ceremonies held in the high school auditorium. That year Meriden Township schools had ten of the graduates.

In the hope of getting more farm boys to improve their education, OHS offered a special agricultural short course for young men age fourteen to twenty-five who did not have an eighth grade certificate. The starting courses were farm arithmetic, English, and business classes. Later, Ellis Clough, head of the school agricultural department, launched a special program for farm boys that consisted of two agricultural courses and two academic subjects.

A Wet Vote and World War I

In July 1915, two medical doctors and an attorney from Owatonna conducted a meeting at Grandprey Hall at which they spoke on the county option—wet or dry. The beer drinkers of the township had their minds made up and voted 166 to 23 to remain wet, and Steele County residents voted 2,508 to 1,599 wet. That was the largest margin of any county in the state. Prohibition came later.

In June 1916, the National Guard unit from Owatonna was called to the Mexican border. This was a sidetrack to the war that was going on in Europe. The June 22, 1917, *Journal* carried a list of all who had registered for a possible call to the service. Eighty-four from the township who were eligible for the draft were listed. Others were added as they became draft age. America declared war in April 1917, and many from the township were called to duty.

In October, the third and fourth draft calls took place at the same time the Second Liberty Loan campaign was conducted. M. S. Alexander, F. H. Joesting, and L. F. Hammel, prominent Owatonna civic leaders, called on people of the village and more than $5,000 was pledged. Soon after the war started, Carl August Drache, my great-grandfather, had to register as an enemy alien. He had been here since 1893 but had never bothered to apply for citizenship. In May 1918, Lloyd Grandprey and Fred Radtke were drafted. Radtke was discharged because of a physical disability and returned to his work in the blacksmith shop. In September eighty from the county were called including eight from Meriden. The national goal was $6 billion for the Fourth Liberty Loan, and the county goal was $960,000; Meriden's was $49,700.

Carl August Drache, my great-father taken February 2, 1918.
This photo was taken because he had to register as an enemy alien.

A May 14, 1918, two cent postcard sent to Paul Drache by Lloyd Grandprey from Camp Wadsworth, Spartansburg, SC.

In October, word was received on August 31 that Hubert Spies had been wounded at the front. In October 1919, Herman Mundt returned after twenty-five months serving in the motor truck division, and in April 1920 ,Frank Dinse received his $150 soldier's bonus check. On May 5, 1923, Max Albert Drache, Jr. died as a result of being gassed at the front. He was survived by his widow, Louise Bartsch, who he had married in June 1922, and his parents, Mr. and Mrs. Carl Drache. He was a nephew of Max Ernest, Rudolph Drache, and Emma Wilker. See the Appendix for those from the township who were involved in WWI.

Rural Free Delivery

As stated in the previous chapter, Rural Free Delivery (RFD) came to the township in 1901. Small town merchants opposed RFD, but it was so popular with the farmers that it could not be stopped. Mail order houses boomed because of all the goods that became available through the benefit of the low-cost parcel post system that went into effect in 1912. In 1915, the average route was thirty miles, and salaries were increased $200. In 1916, the Meriden correspondent reported that carrier Eli Eliason had been delayed, first because of car problems and then tire trouble because of "punctures caused by tacks." In March 1919, villagers and those on the route received one of the seventeen postcards sent by Lloyd Grandprey while he was in a rest area for soldiers in the Pyrenees about ten miles from the Spanish border. In May 1920, the horses of mail carrier William McDonald had a run-a-way. They became frightened when part of the harness broke, and the buggy ran off the road into the blacksmith shop. L. C. Palas, who was planting potatoes in a field along the road, caught the team.

McDonald had the Meriden route for many years. In the winter he always had iron slabs about two-inches thick and about six by eight inches that were laid on Anna Drache's stove. The last thing McDonald did after he had delivered his mail and picked up the mail for the rest of his route was to fetch the cold iron from his sleigh and bring it to the stove and retrieve a hot iron.

People watched for the rural mail carrier every day. I recall when staying with my grandparents, either in Lemond or in Meriden Township, how everyone seemed to know just about when he would pass by their driveway so someone and the dog could make the trip to the mail box. How anxious they were to open the paper and find out the latest news, the *Journal* and/or *Journal- Chronicle, Waseca Herald* or *Waseca Journal* for community news, and the *Daily People's Press* for the business and world news. The weekly paper had the major social news from Owatonna and society, church, farming, road, school, county agricultural events, library notes, obituaries, and miscellaneous mention from the county seat, plus news from the twenty-six community correspondents. All that news was in addition to any news that was gathered from rubber-necking on the party line.

Because many rural roads were still not graveled, substitutes met the carrier at designated places and delivered the mail on those roads. route nine, which ran through Deerfield and Meriden, became part of Route three covered by McDonald. Roads were gradually improved, and each time a mail carrier retired the number of the rural routes was reduced and lengthened. In 1927, the routes out of Owatonna were cut from nine to eight and the mileage was increased from thirty miles to forty. In 1927, the rural routes in the county were increased from eighty miles to ninety-five, in 2002 to 127 miles, and in 2002 to 131 miles. Better roads, better automobiles, and fewer farmers all contributed to the change. In June 1928, the postal service dropped the rate for postcards to one cent because the public objected to the two-cent charge. Postcards were more popular than writing a more formal letter.

Clubs and Related Activities

Virena Stendel Palas was the Meriden correspondent for both the *Waseca Herald* and the *Journal-Chronicle* from 1918 through 1924. She was paid ten cents a column inch and was given a free subscription to each paper. She was instructed to put everything in the paper—card parties, afternoon ladies parties, all club activities, family events, business activities, storms, road conditions, crop reports, township doings, and misdoings.

World War I introduced the airplane, and in September 1919, the Meriden reporter commented how an "aeroplane had circled the village." The excitement was even greater in October 1920 when an "aeroplane" landed on the L.C. Palas pasture just north of the tracks and east of the new school. Alvin Schuldt had a ten-minute ride. This was one of the planes that had appeared at the county fair.

During these years the Steele County Federation of Women's Clubs listed fourteen different clubs that had affiliates in the small communities of the county. Several clubs were mentioned in the previous chapter. Those that persisted were the Larking Club, which met monthly throughout most of the township. A Crochet Club concentrated on needle work and seemed most focused on ladies in the village. There was also a Meriden Sewing Club, which was a mixture of village and rural women and had a different project each year. The Worthwhile Club centered in the eastern portion of the township from Crane Creek and south. Its major focus was household management, furniture decoration, nutrition, and a canning bee—where they canned a different product at each meeting—children's clothing and Christmas candy making. At one meeting the roll call was answered "How can we make our life more useful." At another, a seed and plant exchange took place; another was a program on the cooking of various meats. At its August 1929 meeting it had a report on homemaking and a report from Mrs. C. I. Buxton on her trip to Egypt. All of their meetings ended with a lunch. The group voted not to join the Steele County Federation of Women's Clubs. The Victory Club, which apparently started during WWI, had basket social fund raisers and dancing parties and met in the Meriden Hall. The Cheerful Thirty Club met in homes, and their major interest seemed to be having a good time. At one meeting the debate topic was "Long Hair is More Sanitary that Bobbed Hair." One meeting was a picnic at Beaver Lake for selected boys and girls of Meriden Township, which was project oriented—livestock, poultry sewing, canning—very much like 4-H Clubs. Other townships had similar clubs, and in 1926 the Farm Bureau sponsored the top boys and girls from each club to attend a three-day camp at Beaver Lake with "the maximum of recreation and the minimum of study" under the direction of the county agent and county home agent.

At least seventy-five homes were involved in home management projects and hot lunch programs for the schools. The Meriden Nutrition Club was made up chiefly of women in the village and focused on food needs, essentials of good nutrition, and mother and child feeding, and was assisted by the county home agent. The Wise Owl Club, which centered in the northeastern part of the township, met at homes. Their program for June 1928 was about hogs and closed with a debate, "Resolved That a Hog's Time is Too Valuable to be Spent Husking its Own Corn."

One of the highlights for clubs for the years covered in this chapter took place in June 1928, when 200 women from twenty-five women's clubs of the county gathered in the Medford High School to discuss the formation of a county historical society. Dr. Theodore Blegen, assistant superintendent of the State Historical Society and professor at the university, gave the keynote address, "The Minnesota Historical Society and the Value of Local History."

The first major event in Grandprey Hall after being relocated in July 1915 was conducted by the Stanley Comedy Co., which made a return performance because of its success in 1914. The week-long appearance featured "high class vaudeville and moving pictures and the entertainment is promised to be clean and up-to-date." The usual sponsored dances, basket socials, wedding dances, and traveling road shows continued, but activities in Owatonna and Waseca were steadily becoming more attractive as transportation improved. In January 1916, the CNWR ran special trains from Waseca to Owatonna with stops in Meriden for all who wanted to see the movie *The Birth of A Nation* at the Metropolitan Theatre in Owatonna. Matinee tickets were 25 cents to $1 and evenings 50 cents to $1.50. The highlight for the 1917 Fourth of July was a moving picture under a tent of "several side-splitting-comedies with a feature 'The Battle of Cameron Dam,' a fight between the logging interests and the John Dietz family," which was a big news event of that era.

In the summer of 1926, Howard Haling, the CNWR agent, and Herb Fette started motion picture shows at the Hall every Sunday evening. That was so popular that soon after they gave performances on Wednesday and Sunday evenings. As an added attraction, Haling gave vocal solos accompanied by Miss Hogan at the piano. Mildred Anderson Fette recalled the big event that was held at the Hall on December 10, 1928. She and Herb Fette were married and after they were showered with hay and oats they went up to the Hall where they had a wedding dance. Then they opened gifts after which they had a meal and then danced until 2:30 a.m. That was her introduction to the Meriden community where she played a very welcomed role for the next fourteen years.

In 1920, after the new school was opened, the old schoolhouse was moved next to the bank building and was converted to a hall. One of the first events held there was the nineteenth birthday party for Reuben Luhman. About 100 people attended. The hall was decorated red, white, and blue, and a phonograph provided music for dancing. Supper was served, and after the dancing a midnight lunch was served. There was no further information about events in the hall until the Ahler Brothers were contracted to build an addition to it for living quarters. This then became a pool hall/barbershop, confectionary, and residence for the barber and his family.

In 1921, forty-one individuals put up guarantees for a four-day Chautauqua, which provided an excellent program but because of another event held nearby at the same time the attendance was reduced and each sponsor had to contribute one-third of their guarantee. There was some discussion about having another in 1922, but nothing came of that.

In August 1922, the elevator safe was robbed of $225, $135 in cash and a $100 check made out to Hayes Lucas plus another check for $15 belonging to the elevator. Ernest Schuldt, the yard manager, was out of town, and his helper had put the money in the elevator safe. The robber relocked the safe and the doors when he left.

In September, Paul Drache and Steve Gontarek were hunting near Goose Lake when they heard a call for help. "A Mr. Smith and Mr. Hurdebrick of Waseca had rowed to the middle of the lake and their boat started to leak and sank. Smith could not swim and was under water when Drache and Gontarek arrived. They took him to Waseca where he recovered."

In 1920, the Steele County Fair Board voted that with the new facilities they should have a five-day fair in 1921. That was a good decision ,for the attendance rose to over 32,000. In 1922, Meriden business firms all closed for one afternoon so everyone could attend the fair. The new facilities proved to have an added benefit, as they were adequate to attract large gatherings. In 1924, the Ku Klux Klan had an event at which they professed that they were not "anti anything but were Pro Everything for which the Stars and Stripe Stands." Many from Meriden attended as they did the following year when the Klan had the Memorial Day program at Albert Lea. The Klan's local spokesman, Rev. M. H. Frye, contracted for a meeting for the fall of 1925 for which the Klan paid twenty-cents per person attending and had to furnish bonding for any damage to the property. That September they had 5,000 at their gathering. They came again in 1926 and had a torch-light parade at which they expected to have a crowd of 10,000 but more than 20,000 packed the city streets. The 1927 "Klonklave" had 10,000 turn out, but after that there was no mention of the Klan.

In , the fair board decided that they wanted to have a free fair. The decision was based on the assumption that the fair should function as a public service, and they would rely on income from the concessions to operate. That move was a real spark; attendance jumped to 71,841 up from 40,183 (the previous high). In 1928, there was a heavy rain during the fair, but the attendance rose to 72,081.

The most noticeable change in the township during the years covered by this chapter was that many residents were employed in Owatonna and Waseca. The State School had been an employer of Meriden people for many years, and retail firms provided employment opportunities followed by Federated Mutual, the banks, and numerous manufacturers. The Owatonna Gavel Club, made up of the business leaders, was the prime mover in getting businesses to move to the city and also in getting rural people to seek employment there.

Judging from the above commentary, organized social life accelerated considerably during this era, but there was still much impromptu activity taking place. For example, in the May 25, 1928, issue of the *Journal-Chronicle* the Meriden column had forty-seven entries about people making visits for social events during the previous week. That did not take into account items of two people who purchased new cars, another who broke his wrist cranking a tractor, five who came to the village on business other than shopping, a shipment of a carload of livestock to Winona, two people who came to audit the books at Hayes-Lucas, an announcement of school closing for the summer, and reports of many who went fishing. Does that explain why everyone knew what was taking place?

Meriden was a small but thriving village, and there were people who still saw opportunity there and enjoyed its life style. I remember Sunday, September 23, 1928, when my parents, Paul and Anna Drache, and Mr. and Mrs. O. P. Jensen drove to Albert Lea. When they returned they found ninety relatives and friends gathered at the Drache home all set to surprise them on their fifth wedding anniversary. A 7:00 p.m. supper was served. Rev. Harrer gave a congratulatory speech after which gifts were given and a silver purse was presented to purchase a piece of furniture. That trip was made in the 1926 Dodge, which served my parents for many years. I recall the excitement of my folks when they saw the house completely occupied and the laughter of the Jensens, who were such a jolly good couple. They were so pleased that they could pull the surprise on the folks.

In 1928, a national landmark of the county changed ownership when the National Farmers Bank building in Owatonna was purchased by Don and Mark Alexander for $63,000. This "most noted banking structure in the country," designed by Louis Sullivan and built in 1907 at a cost of $126,000, would continue to be a bank. Security Bank occupied the portion of the building that was designed for banking, and the *Journal Chronicle* used the east portion.

Interest in airplanes continued, and in late 1927 Owatonna held its first discussions about an airport. During 1928, the army promoted flying events to aid the cause of air mail as well as for military use. Fifteen planes were traveling across the country, and in June over 8,000 people came in 2,000 cars, which packed a temporary airport east of the city, "to observe the largest group of aircraft ever assembled here. The fliers thrilled the crowd." In September sixty business firms in Owatonna signed a petition for air mail service. A climactic aerial event took place in August 1929 when the American Legion sponsored "Carnival of the Sky" in which Nona Mallery, the world's premier woman parachute jumper, jumped from a plane that was flying at 2,000 feet.

During the years covered by this chapter the congregation of the German Methodist Evangelical Lutheran church gradually eased out of the picture when the congregation consolidated with the English speaking group in Owatonna. The building stood idle for many years, and in September 1927, L. H. Walther purchased it but did not know how it would be used. That left St. Paul's Lutheran as the lone congregation in the township. The livestock shipping association and the cooperative elevator were the other entities that went out of existence. Other than that, the township and the village were well established in 1929 and probably better prepared to face the decade ahead than many communities of the nation. [i]

---

[1]   Thorstein Veblen, "The Country Town," *The Portable Vebelen*, ed. Max Lerner (New York: Viking Press, 1948) 407–410; Dorothy Wicklow Hudson, letter, 17 November 1999, hereafter Hudson letter; *Journal Chronicle*; Drache, *Furrow* 24, 52; Virena Stendel Palas, personal interview, 14 June 1996, hereafter Palas interview; Gertrude Hythecker Rietforts, letter, 10 May 1989; Meriden RTC; Drache interview; R. M. Hugdal, Acting Inspector in Charge, letter to Postmaster Meriden, Minnesota, 28 December 1918; Grandprey Jr. interview; First State Bank of Meriden, Minutes 1915–1959, hereafter Bank Minutes; Mildred Anderson Fette, personal interview, 14 June 1996 and cassette 10 March 1996, hereafter Mildred Fette interview; Peggy Korsmo Kennon, transcripts of Dick Rietforts for Farm America, 8 January 1986, hereafter Kennon transcript; Donald Born, personal interview, 8 September 2008, hereafter Born interview; E. Dana Durand, *Cooperative Livestock Shipping Associations in Minnesota, University of Minnesota Agricultural Experiment Station Bulletin No.* 156 (St. Paul, 1916); Paul Buecksler, personal interview, 8 July 1989, hereafter Buecksler interview; Hilda Willert, personal interview, 11 September 1976, hereafter Hilda Willert interview; Drache, *History* 268–269; Audit Reports, 31 May 1924, 31 July 1926, of Meriden Farmers Elevator and Mercantile Company, prepared by Charles H. Preston, certified public accountants, Minneapolis; Hiram M. Drache, *Plowshares to Printouts* (Danville: Interstate Printers & Publishers, 1985) 9–10, 35–36, 59ff, hereafter Drache, *Plowshares*; Robert Scholljegerdes, letter, 29 January, 2002. Successful Farming, "Greatest Tractors of All Time," vol. I, 8–9, 14–17; Gertrude Scholljegerdes Voss, personal interview, 16 July 2002, hereafter Voss interview.

CHAPTER VI
Maturation 1930–1944

The Chicago & Northwestern

In June 1930, the CNWR advertised an excursion to Milwaukee for $6.45 and to Chicago for $7.55 round trip, no baggage checked. There was no further comment, but in July passenger service from Winona west was reduced to three trains each way daily. These were all stop trains, so the postmaster was required to be at the depot for each train. Then, in September 1931, the road was granted permission to drop one train each way, which would save $30,000 annually. But traffic continued to remain depressed, and in August 1932 none of the trains stopped except when notified in advance.

In 1930, the Minnesota division had been combined with the Dakota division. Elmer Johnson was laid off from his job, but because of seniority he was able to land the Meriden job in March 1931. Johnson was pleased, for he earned $125 a month and had no debt. The family was thrilled when they arrived in Meriden and found that they would be living in one of the finest houses in town, and the rent was only $12 a month. They were disappointed in 1934 when my maternal grandparents, Mr. and Mrs. W. R. Schultz, who owned that house, retired and moved to Meriden. The Johnsons had to move and the only place available was above the Luhman store, which had not been remodeled since 1879 when it was the village hotel.

The section crews were reduced when the divisions were combined ,which left Floyd Walters, the Meriden foreman, as the only crew member, but within a month the men were hired back half time. The Meriden and Owatonna sections were combined, and the Meriden crew was in charge of the enlarged section one week and the Owatonna crew the next. By August 1932, Meriden no longer had any train that made regular stops and only one would stop on signal for passengers, parcel post, or express. In 1933, all farmers and railroad section workers were engaged to spend the week of July third eliminating weeds to avoid further damage to crops.

But not all was gloom, for when the Chicago World's Fair opened in 1933 the CNWR offered a ten-day round trip coach fare for $7.55, or a sixteen-day round trip berth and parlor car for $14.95. In each case, street-car or motor coach were available every few minutes to the fair grounds and return. During the February blizzard in 1936, the entire CNWR system was shut down for three days before the snowplows could get through.

In desperation to maintain passenger service, the road introduced two "Minnesota 400" trains (400 miles in 400 minutes), one from the Twin Cities and the other from Mankato, to Milwaukee and Chicago. The trains had air-conditioned coaches and a café lounge car for $13.58 for the coaches and $17.10 for parlor car from Owatonna. The "400s" made limited stops, but Waseca and Owatonna were stop stations. I remember how thrilled I was when I saw the first "400" go through Meriden at sixty miles per hour and felt the street by our house shake. Despite these valiant efforts and the weak economy, the railroads could not overcome the fact that people preferred traveling by car.

In 1931, a natural gas pipeline from the southwestern part of the country was laid to the Twin Cities. The line ran from section thirty-three in the south through four in the north with a branch line to the east to Owatonna and Rochester. The pipes were sixteen-inches in diameter and from twenty-four to forty-eight feet in length. They unloaded in the vacant lot between the stockyard and Hayes Lucas. Thirty-seven carloads, each carrying thirty-six pipes, were shipped in. We Meriden boys had exciting days watching the hoist load the big semi tractor pole-trailer rigs and then listening to them "growl" through the mud to get on the road.

In 1932, for reasons of economy, the CNWR decided to assemble an entire train of sixty-five cars of sugar beets in the Waseca yards for shipment to Mason City. Each day the cars that were loaded in Meriden were taken by a switch engine to Waseca and then empty cars were placed on the track to be loaded the following day. That is an early concept of what became the uni-train system that favored stations with longer sidings and larger facilities.

The railroads continued their struggle, and in December 1937 the Interstate Commerce Commission announced that the 71,386-mile railroad was in equity and bankruptcy court and many more railroads were on the brink of failure. At 11:00 a. m. on April 4, 1939, W. P. Jones, Lloyd Grandprey, and the chairman of the telegraphers' union appeared in the depot in Meriden before George Dames, attorney for the Railroad and Warehouse Commission, to hear why the road should continue service. From 1936-1938, total freight revenue was $19,537.80; passenger revenue, $914.50; lease revenue, $151.94; and mail revenue $356.21.

After Elmer Johnson was transferred to Winona in May 1935, the road went to a part-time, non-telegrapher agent six hours per day. "The population of the village was about 100 and there had been only nine industries forwarding or receiving carload or less than carload shipments, since 1936." Permission was granted, and Lloyd Grandprey was appointed to maintain custodial service of the station. Hayes Lucas was the largest customer during the years cited, and Grandprey formed that duty until the Hayes Lucas closed in 1963.

On September 27, 1939, the "400 streamliner" with a diesel engine capable of pulling a full load at 117 miles per hour was debuted. The fuel it burned cost eight to nine cents a gallon and it used 50 percent of the heat it produced. It had a cab on each end of the train so it did not need a turntable like the steam locomotives and was much more efficient to operate.

The only other activity of the CNWR that attracted any attention came in 1937 when seven cattle owned by Ewald Lewer were killed by the eastbound evening passenger train. In 1940, a locomotive tire on train No. 514 broke near Meriden on June 29 and held up the traffic for two and one-half hours until they were able to get on the siding.

In January 1942, the company inaugurated a new Minnesota 400 to operate between Mankato and Wyeville, Wisconsin, near Tomah, here the two rail lines crossed, to "relieve other passenger trains for wartime requirements." Those of us who rode the passenger trains during World War II appreciate what the railroad did in that era. Many older passenger cars were brought into service, but they were always filled to capacity. The final daily contact with the railroad took place during WWII when the postal system switched to trucks for delivering the mail.

Meriden Businesses

Telephones

During the 1920s, the two grocery stores, Fette Garage, Hayes Lucas, and the bank had both the Waseca and Owatonna telephones installed. By 1930 telephone usage had grown sufficiently so two linemen were needed to assure good service—one for the eastern area, which was referred to as the Crown Rural Telephone Company, and one for the western area called the Meriden Rural Telephone Company MRTC. By October 1931, six village businesses had both Owatonna and Waseca phones. Only the Palas Garage, the blacksmith shop, the barber shop, the creamery, and the elevator did not have a phone at that time.

In 1932, the Tri-State Telephone system, which controlled the access to the Bell system, stated that all the rural lines had to be in good condition if members wanted to switch to dial phones. The village subscribers must have been content with the traditional wall phones, for it was not until the late 1940s that the village had dial phones. In 1942, the annual phone rent for the Owatonna phone was $15 and $18 for Waseca, plus the charge for long distance, but for most residents in that era long distance calls were a rarity.

In 1925, the MRTC took out liability insurance, which proved to be a fortunate move; on June 3, 1931, Mike Ebeling, the lineman, met his death while working on the line. In January 1933, Mrs. Ebeling was awarded $7,500 in addition to $500 funeral and medical expense. That year renters were charged $12 for the year and the stockholders paid no annual rental but were assessed $8. In 1934, the annual rent was increased to $15. The stockholders did not have to pay the rental charge, but they were assessed $10 for upgrading the line. In 1935, some members requested to have an additional phone in an out-building, usually the shop or milk house. In 1938, the decision was made to incorporate. New members had to purchase stock at $75 per share to finance improvement of the line and for telephone boxes. Current members were assessed $8. In 1939, the creamery was connected. In 1941, a 10 percent tax was added to all assessments and rentals

Postal Activities

In June 1940, William J. McDonald, who had used buggies, sleighs, Model T Fords, and finally a Model A Ford equipped with snow traction gear, ended his thirty-year career. He had served nearly all of those years on rural route three and later on route one. He endeared himself to many because of his dedicated and punctual service to township residents. He was replaced by Walter White.

On November 10, 1942, the mail pouch that was tossed from the train was sucked under the train. The contents were damaged and widely scattered along the right-of-way from the depot and the east end of the siding. The pouch, which was intact, contained $15,000 in cash from the Federal Reserve Bank to the Meriden bank. Anna Drache and John Evans, the section foreman, delivered the contents to the Waseca postmaster. The next day members of the section crew found more letters and parcels but apparently everyone, including the bank, got the mail they were expecting, for no requests were made.

Barber Shop and Meeting Places

Harvey and Grace Moore had operated the barber shop, pool hall, slot machine, confectionary (including tobacco products and pipes), and a lunch business since 1924. They had developed a very lively, well-mannered community meeting spot, comfortable for all. They accommodated everyone from a child who wanted a cone of ice cream to a grandpa who wanted his beard trimmed. The barber shop area was semi-enclosed with a wall just high enough so that adults could lean on it while they carried on a conversation with Harvey or, more likely, the person getting a shave and/or a haircut. The conversation varied from the local robbery, a bit of gossip, or business. At times when Harvey had customers standing in line, he would lower the over-hanging glass panels to shut out the visitors and the commotion from the adjacent pool tables.

The business opened every morning at eight or nine depending on the farm season and was nearly always open until 10:00 p.m. The Moores lived in the attached residence, so the business virtually consumed their life. They needed a break, and in September 1930, they sold the business to William and Rose Arndt and moved to Faribault, where Harvey worked as a barber and Grace worked as a store clerk. The Arndt family moved into the living quarters, but the children were very seldom seen in the business area. The building was still owned by Fred Fette.

The grocery stores both sold candy, ice cream, soda pop, and tobacco. If the store was not busy, the grocer often had to wait as the children pondered over whether to spend a penny for a Tootsie Roll or a nickel for a Baby Ruth bar. After 1926, gasoline, confectionary goods, and sometimes lunch were available at the station at the junction of the county road and the highway. Various members of the Herman Brase family—Harold, Ernest, or Herman—operated it, so originally it was called "Brase's Place" or simply "The Corner," and finally "The Midway." At times, under-the-counter liquid refreshment was available to give the near beer extra zip.

Originally, there were no living quarters at that location so it was a good spot for robbery. On the first Sunday in May 1932 the station was robbed of a "few bottles of soda pop, some candy bars, and the desk drawers were rifled." On October 26, 1932, the Arndt family was away from the barber shop/pool hall and Herman Larson, a long-time resident of Deerfield/Meriden, was in charge. The business was robbed of $64 at about 11:00 p.m. by "a trio who posed as federal officers on a Prohibition raid." A week later three Owatonna men were jailed for the holdup.

Things changed in 1933 when Prohibition was repealed. Owatonna announced that it had granted seventeen on-sale and thirty-three off-sale beer licenses. That April, on-sale licenses were granted to William Arndt, H. H Wicklow, H. J. Luhman, and Ernest Brase, and an off-sale license to H. H. Wicklow. I recall the morning on my way to school seeing a new, brightly painted beer truck backed up to the pool hall steps and cases of beer being carted from the truck.

My paternal grandfather always made home brew and also made root beer. My maternal grandfather was a teetotaler, but he had a table spoon of Alpenkrauter three times daily. Alpenkrauter was a patent medicine that came in an attractive square bottle with raised letters in German on all four sides. It was similar to Kuriko, which the Norwegians used for their ailments; it came in a round bottle. Our family doctor told us that they both had an alcohol content of about 13 percent prior to Prohibition, and when Prohibition went into effect it was raised to about 30 percent.

In October 1933, the Brase station and lunch room, then operated by Mike Gontarek, was destroyed by fire but was rebuilt and then operated by Herman Brase. In May 1934, Harvey and Grace Moore returned to Meriden and operated the barber shop and related business until 1942 when they retired in Owatonna. In 1939, Herman Brase, Harvey Moore, and Henry Wicklow ran the three places in the township to have on-sale and off-sale licenses.

General Merchandise

In February 1932, while the Wicklows were attending a dance at the hall, a thief gained entrance to the store via a stairway to the basement. Much of the inventory was kept there in addition to what was stored in an outside shed. "The thief stole two boxes of gum but nothing else was touched although there was a considerable amount of money in the cash drawer."

In April 1935, Homer C. Donaker, owner of the Self Serve Store, a chain of six stores headquartered in Owatonna, purchased the Luhman store. H. J. Luhman had operated the store since 1898, and from 1917 to 1928 he was postmaster. During most years he had rental income from upstairs apartments. He made an adequate living and wanted to retire. Edward and Ellen Spiekerman, who had fourteen years experience in the grocery business, were engaged to manage the store. Donaker stated that the chain had a large volume and would provide lower prices. Each week they placed a large ad in the *Journal*. The new store introduced more updated techniques, but Wicklow remained a strong competitor, for, like the Moores, the wicklows were a real "mom and pop" team. Wicklow's daughter commented that her mother told her she took a few weeks off when Dorothy was born and a relative took her place in the store. Theirs was truly a general store offering yard goods, overalls, shirts, shoes, a limited line of drug items, and a full line of groceries, tobacco goods, beer, soda, and candy. They purchased eggs and live chickens for cash or trade. Much of my early clothing, including shoes, came from that store. The store was open every day of the week except Sundays from the time the first teams came to the creamery until at least six in the evening and as late as ten on Saturday nights. The Wicklows lived on the second floor, so it was not uncommon for customers to come after hours or even on Sundays. Wicklow joined the Blooming Prairie egg pool, which merged with the Southern Minnesota Grocers Association, in an effort to get better prices for poultry products and to buy groceries cheaper so they could compete with the chains.

In 1930, Wicklow became a dealer for Central Cooperative of Owatonna and installed pumps for Hi and Low test gasoline. Dorothy Wicklow Hudson Photo.

First State Bank

At the first bank meeting in January 1930, the directors learned that the bank had a net earning in 1929 of $6,852.50, and they declared a 10 percent dividend. The other order of business was to revise the by-laws to raise its liabilities to $1 million up from $450,000. But all was not well. Fred Fette's handwriting had showed signs of decline since 1927, and in recent months his health had declined. In May 1930, he resigned as cashier after fifteen years in leading the bank and being a real spark plug in several other businesses in the village. At the June 24, 1930, board meeting the bank examiners produced a long list of criticisms that were apparently the result of Fette's failing health and were easily corrected. Ewald Fette, who had been assistant cashier, was named cashier at a salary of $200 a month.

At 10:30 a.m. on October 14, 1930, assistant cashier Alfred Schuldt and W. E. Galloway, a bank customer, were accosted by two gunmen with blackened faces. In less than five minutes the robbers scooped up $1,689 in cash and then forced Schuldt and Galloway into the vault and left. "A number of residents realized what was happening." The vault was not properly locked so Schuldt was able to work his way out. Only four dollars on a cash tray was left behind. On October 26, two of the four men involved in the robbery were caught, and on November 7, one was sentenced to life in prison. Another was wanted for questioning in six other bank robberies.

On November 1, 1931, W. P. Jones was named cashier of the bank, but because of the housing shortage the family was not able to live in the village until June 1932 when they moved into the Ewald Fette home. The paper reported, "Steele County banks were more than remaining firm.... [They] have shown gains which have been surprising in comparison to conditions elsewhere....The County is fortunate compared to other agricultural areas and is also gaining an edge on industrial areas where unemployment and reduced incomes exist....Any improvement will depend largely on an increase in prices received by the farmers."

Jones was quoted, "The people will only buy what they can pay for, thus providing more healthy business conditions. The bank had a very good year, despite conditions last year, gaining materially in the number of depositors and customers." During the Bank Holiday of 1933, all Steele County banks took immediate advantage of federal orders allowing the two Owatonna banks, Blooming Prairie, Ellendale, Medford, Meriden, Hope, and Bixby to reopen and return to business by March 1.

Jones took charge, and bank assets rose and declined depending on weather or political events that affected agriculture. All the while he strengthened his position by steadily purchasing shares of those who, for various reasons, were willing to sell them. By 1938, there were only thirty-eight stockholders remaining, and Jones had regained majority control.

In November 1943, Jones wrote to the Commissioner of Banks and asked for permission to declare a $30 per share dividend. He stated that in the recent examination there was not "one item of classified loans or other items...[we] have a pretty clean slate." Within three days F. A. Amundson, Commissioner of Banks, replied that Jones had one share of stock in 1932 and now he owned 116 shares. "This meant that he would receive $3,480 of the total $7,500 to be disbursed." The commissioner continued that he had no record that Jones had made any contribution at the time of the reorganization. Then he asked if it was wise to reduce capital funds at a time that the bank was having rapidly rising deposits. Jones replied the next day asking if he had done anything wrong and stated that he "had given the twelve best years of his life, seeing that the depositors were repaid every dollar of their deposits [lost in 1933] and I had not received one cent of compensation. I was in no way to blame for the condition the bank was in when I took over in 1931, which was a difficult job....When I came here, I realized that if I were successful in my operation of this bank, the only way I would be compensated...would be through the acquiring of stock and increasing its value and earnings."

The bank placed an extensive article in the *Daily Peoples Press* stating that the bank had repaid the $152,625.84 in trust deposits as established by the government following the 1933 bank holiday and that Jones and H. H. Wicklow served as trustees without pay to get that job accomplished. Jones went on to justify his position and wanting to declare a 30 percent dividend. The commission yielded to the request for the dividend; then Jones immediately asked for permission to declare a 20 percent stock dividend, which the commission denied without giving reason. The entire Meriden board sent a resolution stating that they had legal advice that they could declare a stock dividend.

In January 1944, the bank sent a letter to all checking account customers that they would be charged for checking privileges. A $100 deposit entitled a customer four checks per month. In June 1944, Jones wrote to the Federal Reserve seeking advice as to whether or not Meriden should convert to a national bank, and they replied that there was no advantage or disadvantage. For an accounting of the dividend payments of the bank, see the Appendix.

Hayes Lucas

On October 20, 1931, thieves broke into the Hayes Lucas through a rear window and stole about $40 worth of jack knives, razors, and flashlights. This was the third burglary of the year following one on August 25 when the depot was robbed and on September 3 when someone broke into the pool hall/barber shop. Thieves returned to Hayes Lucas in June 1933 when they pried open a window in the office and stole jack knives and fishing tackle worth about $40. On the same night they pried open the ticket window at the depot and stole candy from the vending machine. Police and railroad detectives investigated. I have fond memories of those little Chiclets packs of hard chocolate and chewing gum that were dispensed from those machines.

In 1933, a Drache truck hauled a load of hardware goods from the Hayes Lucas at Douglas, Minnesota, near Rochester, which was being closed. The Meriden store continued to do steady business, and in 1936 electric lights were installed. I have memories of Lloyd Grandprey working nights by the light of a mantle lamp that hung over his table in the store. A space heater served as the separation from his work table and the display tables and shelves that held everything from kitchen items to harness.

Herb Fette's Enterprises and Other Garages

Herb and his brother Ernest (Ernie) Fette ran the garage from 1923 until 1929 when Herb suffered a seizure and the doctor told him to get out of the garage for a while. He was not one to sit and took an offer to operate a movie theatre in Shakopee. Ernie then ran the garage until the highway was being built in 1930. The road contractor made an attractive rental offer for the garage for servicing their equipment and for the house as a dining hall for the workers. After the construction was completed, Ernie reopened the garage and carried a full line of accessories and supplies needed for auto repair. In 1932, Herb returned to Meriden to resume the business.

The only news about the Harvey Palas garage for 1932 came in April when thieves cut the gasoline hose on the pump at the garage and drained the glass reservoir of its supply. The maximum the reservoir held was ten gallons which meant that the loss in gas would probably have been about $1.80. A new hose would have been twice that amount.

After Herb Fette returned to Meriden, he started working on what his wife said he had long dreamed about—a radio station. Initially, Herb was so focused on the radio station that he was not attending to the garage business, so they had little income. Deichen threatened to cut off their lights, so Herb hooked up a Model T engine and made his own generator that could charge batteries in four hours to keep lights in the house and the garage for a day. Work on the radio station progressed, and Millie, aided by Ernie's wife, Evelyn, wrapped wire around oatmeal boxes to make coils for the station. Then, two towers about fifty feet high were erected on the northwest corner of the garage and the southeast corner of the pool hall, which the Fettes still owned, to make an aerial for the station. I remember that those towers remained in place for many years to remind everyone of "what might have been." The call letters were H2RB, which had a wave length of 930 kilocycles and a broadcasting radius of about fifty miles. In January 1932, H2RB was allowed to broadcast only from 11:00 to 11:15 a.m. and from 8:15 to 10:00 p.m. Later a special midnight test program carried the Muskrat Twins, of Silver Lake, and Jankes German band, of Owatonna. In 1933, they were switched to 955 kilocycles and given three one-hour slots daily plus a three-hour period Sunday afternoon. The *Journal* stated that the local people enjoyed the broadcasting and called in to dedicate numbers to their friends. On Sunday afternoon the Victorian Orchestra gave one hour of dance music followed by numbers by some Waldorf musicians who played a Hawaiian guitar and another played a banjo. This was followed by a program of western music.

Herb had no trouble getting the bands that played for dances to appear on the program, for it was free advertising for them. The big problem was that he could not get advertisers because they thought that radio advertising would not pay. He was a few years ahead of his time in that respect. I remember all the excitement this caused in the village and seeing the different entertainers who came to do their broadcasts. Everyone was excited about doing what they could to make it work, but when Herb went to Washington to get final approval for a license, he learned that station KSTP was opposed to the project. He lacked the funds needed to fight for a final approval, and the village radio station was closed down. All Fette had for his effort was a suitcase full of letters from the people who had heard the programs.

Herb Fette and his portable feed mill, which he operated during the 1930s when the elevator was not open.
Barbara Fette Anderson photo.

While the above activity was taking place, the Fette garage contracted with the Sinclair Gas Company in 1932 to install two pumps (Hi test and Low test) on the busiest corner of Meriden. They remained there for three or four decades. Millie Fette donned corduroy pants and sold gas, oil, and minor parts to customers. She also took calls for the operator of the truck-mounted portable feed mill, which was profitable during the years when the elevator was closed. He also had a bulk truck to deliver gasoline, kerosene, oil, and grease to customers. But Herb liked the communication/electrical activity more than those businesses, and in 1933 he started an outdoor movie and public address system business. Outdoor movies were provided to Hartland, New Richland, Otisco, Meriden, Medford, Hope, Pratt, Dodge Center, and the CZBJ Park. Movies were presented once a week from May through September. Millie sold advertising to sponsors who paid $20 to $25 for each show. On show night many people sat on logs or benches and others sat in their cars to view the movie. When the show was over the 9 x 12 foot screen was folded and placed on top of a 1933 Studebaker car, which pulled an 8 x 10 foot projector booth trailer.

The first amplifying service sound truck in 1930 that led to a successful business over both Steele and Waseca countries. Barbara Fette Anderson photo.

The sound trucks were kept busy because few places had amplification systems for football and baseball games, county fairs, political events, and other large gatherings. The most unusual events that the sound truck was engaged for took place at the prisoner of war labor camps during WWII, where they played German musical records. The fee for most events was $20.

In 1932, Ernie Fette opened an electrical service shop in the barn on the Fred Fette home place and was very busy because, like Herb, he was "a natural in electrical work." Herb and Ernie had a private telephone line between Herb's garage and the barn. When Herb moved to Owatonna, Ernie built a tarpaper-covered building on the lot west of the Fred Fette home next to the post office. He remained in that building until about 1943 when he was employed in Owatonna and sold it to Murrel and Isabel Kleist, who converted it to a hamburger shop and pool hall.

In 1936, Walter Klamm purchased and remodeled the old German

Methodist church building in the village for use as a garage and built an addition for storage and an office. The three village garages each had two gasoline pumps, both stores had two gas pumps, and Drache truck line had one, so for a few years the village had a total of eleven pumps, plus two more at the corner station.

In November 1942, there was a big farewell party for Herb and Millie Fette, two people who had been so active in the village. Like the other business people in the village, they had worked long hours to make their business succeed, but they realized that they could not compete with the larger, better equipped garages and service stations. They were both employed by the Roxy Theater in Owatonna, Herb as the projection operator and Millie as ticket seller. In addition, Herb maintained an electrical shop, which had far more opportunity for business than in Meriden. It grew to be a thriving business. The departure of Herb and then Ernie Fette came before the village reached its peak business wise, but they were the first of the original businesses to leave and were not replaced. The garage was sold to Paul Drache for $3,700, who needed it to service his trucks, but he offered service to the public if the mechanics were not busy on trucks.

Blacksmiths and Well Drillers

Blacksmiths were critical to the early farming communities, so they could be particular about where they settled. The village was fortunate that, starting with Peter Pump in 1870, then under the leadership of Sam Grandprey who had as many as four helpers in the peak of the pre-automotive era, Meriden was relatively well served. In 1919, the village had two blacksmiths—Julius Meitzner and Fred Radtke. Radtke left in the 1920s for Cresbard, South Dakota, northwest of Aberdeen. In November 1930, he had Paul Drache move his furniture and blacksmith equipment to Owatonna, where he opened his shop on North Cedar Street and which he operated until he retired in the late 1950s.

As stated in the previous chapter, Meitzner was the only blacksmith in the village for a few years. Meriden was fortunate that Dick Rietforts was available. Rietforts was "all business." He was a hard worker and did excellent work. I spent many hours in his shop during the summers of the 1930s, first as an interested observer of what he was doing, and from 1937 on I helped when he and others built truck boxes and did repairs, under his supervision, on Drache trucks. His wife brought him a fresh thermos of coffee each morning and afternoon, but other than that he worked non-stop. His body carried no excess weight. His daughter, Gladys, recalled that in his later years the family doctor instructed him to take a full hour off at noon and to take a nap. In the afternoon he should go to the house and have a sandwich and a bottle of beer because he needed the calories. He seldom employed any help, but many people came to the shop with projects and worked with him because he was so good at improvising.

As stated earlier, Albert Born also came to the village in 1928. He had a well drilling rig, but it was mortgaged and the banker would not let him take it out of Blue Earth County, so he was not able to work at what he was trained to do. Originally, he could not get help from First State Bank because he was "unknown," but after 1930 when W. P. Jones became cashier, he went to Born and agreed to loan him money so he could pay off the Good Thunder banker and bring his rig to Meriden.

Business was slow at first, but Loren Luhman, the butter maker from 1930 to 1942, asked him to drill a well for the creamery, and he got the break he needed. Born drilled an eight-inch well that pumped 100 gallons a minute. Born's son, Donald, said, "Luhman bragged to the other butter makers about the great well Albert Born had drilled for the Meriden creamery and we got plenty of work." Donald started OHS in the fall of 1935 and exchanged rides with Sidney Brase, who had a Model A, and Johnny Tuthill who had a Dodge. In the second year Donald's dad bought him a "well used" Chevrolet, but by the time Donald was a senior the well drilling business was doing very well. Don and I were boyhood friends in Meriden but Don, in his eighty-sixth year, reminded me, "I had to quit, you know there were seven kids besides me and Dad needed me." He was proud of what he had done, for it was the start of an extremely successful business.

Elevator

After Art Willert purchased the elevator in 1927 he attempted to make it a viable business but did not succeed. Willert opened it in February 1932 and then became ill, so he employed Herman Stendel to do grinding, mixing, and selling feed while he was unable to work. Sometime during 1932, Willert leased the business to Herman Thordson, who had an elevator elsewhere, and he operated it until February 1933 when he gave up the lease and left town. Willert kept the business open under Stendel, who provided a full line of feed and grinding. In a phone conversation with his son, Charles, he said, "My parents never wanted to talk about the elevator. For some reason it was a sore spot for them."

Author's Note: "The building stood idle for some years and Donald Born and I were able to crawl in and take the hand lift to the top, just to get a better view of Meriden. It was very dusty, but there was no evidence of birds or rats. We never gave any thought to what might have happened if the ropes on the lift would have failed."

In 1934, Paul Drache rented the feed warehouse. Drache sold linseed oil meal, bran, middlings, tankage, meat scraps, oyster shells, and concentrate. Farmers needed these feeds to supplement their home-grown grains. They had the Herb Fette portable feed mill do grinding on the farm and mixed their own feed. In those years, Drache hauled livestock to Newport, So. St. Paul, Albert Lea, and Austin. Much of the time if he did not have a return load he purchased a load of feed. Tankage and meat scraps were available at the packing plants, and the other feed came from the Minneapolis mills. He did the same when he had a load of freight to Mankato and picked up feed at the Hubbard Mill. In this manner he was virtually able to guarantee a return load because the only other freight that he hauled at that time was for Hayes Lucas, the blacksmith, the creamery, and the grocery stores. The farmers called or came to our door, and when I was home it was my job to take the keys and go to the elevator. In those days the concentrate sacks were made of printed cotton with various designs. The farmer had instructions from his wife to look for a certain pattern, and sometimes he even had a sample along. To find the correct pattern meant sorting through the 100-pound sacks. The emptied sacks were used for aprons, dresses, other clothing, and dish towels. To prove how much this meant in those days, the Cargill Company came out with patterned sacks and I learned from them that it was probably the biggest sales success that they had ever experienced.

In August 1937, the Olson Brothers, who owned the elevator at Dodge Center, purchased the elevator from Willert and immediately remodeled and up-graded the facility. The twin-beamed hand-crank wagon lift was removed, a larger dump pit was installed, and an enlarged belt and cup system and a truck scale and hoist completed the system for faster grain handling. Henry Olson approached Anna Drache about providing rooms for himself and his son, Buddy. Then he requested three meals a day for himself and Buddy and six crew members for which he paid $0.90 a day for each man and $0.50 for Buddy. They were there from mid-August until the third week in November.

On November 15, 1937, the Meriden Grain Company opened for business under manager Henry W. Olson, and once again the village had a full-scale elevator. The structure had been completely repaired and remodeled, and the new elevator purchased grain, did complete grinding and mixing of feed, and sold a complete line of feed and seeds, plus salt and twine. In January 1938, the company erected a corn crib so they could purchase ear corn but also added a cob shed and grinder. They purchased corn cobs ,which they ground and sold as litter for use in chicken houses. Meriden

Grain Company was a strong contributor to the business community. Olson was a good marketer and responded to the needs of the farmers.

Truckers

By 1928, Paul Drache's health had recovered sufficiently from surgery so he decided to get involved in business because he enjoyed dealing with people. In the previous chapter he had taken over the milk route for the village. He knew that the local creameries would not take sour cream ,so there was always a supply available. He contracted with the R. E. Cobb Company, of St. Paul, which served as a collector of eggs, sour cream, and hides and furs for further processors. He built a garage/cream station on the same lot as his home and installed a cream tester, egg candling equipment, and scales. Farmers home butchered their cattle and had the hides to sell as well as those from calves and horses that died. Hunting and trapping were still very much in vogue, so pelts and fur were available. Hides had to be salted, and furs and pelts had to be stretched. His busiest days were Friday and Saturday because he also bought sweet cream. Because he paid cash, farmers who were creamery patrons but were short of cash for the weekend knew where to get it. In 1929, he purchased a truck to deliver his goods to the assembly point in Owatonna and secured a 1925 REO Speed wagon with an eight-foot rack for $150 from Ed. Newman. Newman farmed north of Meriden, and since he had discontinued raising sugar beets he felt that he had no further use for a truck.

Business grew rapidly, and Drache could not keep up with the REO. Within a few months he purchased a used Model A Ford truck. On many days they hauled two loads of livestock to market in addition to local hauling. Herman Larson, whose parents farmed in Deerfield, was employed full time hauling to Hormel, in Austin, or Wilson, in Albert Lea, as farmers began to feel more comfortable with direct selling. But those who raised beef preferred selling to So. St. Paul. Within the next couple of years the CNWR dismantled the stockyard and the water tower and plugged the well.

In 1931, Drache purchased a new, larger International Model A-3 truck for $550 and had Rietforts build a twelve-foot rack for $100. He soon found himself in the moving business because people saw all the advantages of using a truck instead of the railroad. In 1932, Drache employed Walter Klamm, who came from South Dakota to seek work. He was paid $10 a week and room and board plus expenses on the road. Moving jobs led to traveling in other states, and on return he hauled brick, tile, and cement from Mason City for Hayes Lucas or else he purchased fruit, if in Michigan, or potatoes, lumber, or cedar fence posts in northern Minnesota or Wiscon-

A 1931 $550 IHC truck, which had a top speed of 35 mph. Many people saw the advantage of moving by truck. Paul Drache had a canvas tarp that fit over the rack for hauling furniture. He had moving jobs from Cresbard, SD; Toledo, OH; and Oklahoma before all the roads were paved.

sin. When farmers realized how much easier and more economical it was to get their tile delivered directly to their fields from the tile plant, they were quick to ask for that service.

In 1933, Drache hauled his first canned goods from the Owatonna Canning Company OCCO to Wisconsin, which led to hauling to other states. Soon OCCO became his largest customer. In 1934, he purchased another larger International with a sixteen-foot rack with a ten-foot double deck. I remember seeing farmers standing in awe looking at and talking about "such a large truck." At that time he installed an underground storage tank and purchased gasoline for a net of fifteen cents a gallon, which included four cents for taxes. Many times on Sundays people would drive up to the pump to buy six gallons for a dollar. Dad told them to help themselves, and some people "gave the handle an extra stroke for just a bit more."

In 1936, his largest load was forty-nine hogs, which weighed 11,661 pounds and brought the four farmers $1,060.97 after $17.64 had been deducted for trucking. Drache had a good eye for livestock, and farmers asked him to buy feeder cattle or feeder pigs at the yards that they could fatten. Often farmers wanted to sell sick or thin cattle or boar pigs. If he felt he could heal the cattle he would take the risk and buy them in hopes he could fatten them. He castrated the boars, fattened them, and sold them as stags, which brought a higher price. The largest single shipment for 1936, which came from Henry Rosenau, consisted of twenty-four steers that weighed 26,240 pounds. That was the top of the market at $9.00 per hundred and produced a net check of $2,098.76.

After the Meriden Grain Company purchased the elevator business, he stopped dealing in feed, and they became a regular trucking customer. He was not one to solicit business, but as hauling for OCCO increased to the Twin Cities, Owatonna manufacturers, beer wholesalers, and Central Cooperative Oil Association had loads of batteries, drums of oil, and grease for a return load.

In 1937, Carl Dahlstrom arrived in Owatonna and leased the idle gas plant building with the intention of opening a meat locker cold storage plant. He also established the Owatonna Milk Processing Company. Loren Luhman was a county leader in the creamery circles and introduced Drache to Dahlstrom, who offered him the contract to haul buttermilk and/or skim milk from area creameries to the plant. In 1942, Luhman left Meriden and became the manager of the Borden plant.

In 1939, the trucking business virtually exploded, as the European and Asian wars created a demand that impacted our nation. Farmers responded to the increased prices and produced to satisfy the demand. Dad had watched the changing trends in livestock marketing during the thirties, and he told me that he was amazed how many farmers still preferred the public markets rather than selling direct to Cudahy, Hormel, or Wilson. Field men for the commission firms came regularly to solicit business, but it was not until 1938 that a Hormel buyer called on him. In 1939, Hormel sponsored a tour of the plant and a dinner for 130 farmers who were regular customers of Dad. Most of these farmers were within a ten-mile radius of Meriden. This took place March 28, 1939. It was a tremendous success.

The wartime activity abroad stimulated a great increase in beef and pork production, and it was not unusual for Drache to have two to four loads daily just to Austin or Albert Lea. At the same time, Rev. Bill Robertson, of Owatonna, who worked with a boy's club, heeded the demand for recycled paper and led drives that produced volumes of baled paper that needed to be hauled to the Waldorf Paper plant in the Twin Cities. This offered loads to the Twin Cities that replaced the void in livestock shipments to the Union Stock Yards because more livestock was being sent to Austin and Albert Lea.

By 1941, four trucks were hauling milk and another freight truck was needed just to load out the powdered milk. By 1942, Drache trucks hauled 1,082,555 pounds of livestock, 1,026,357 pounds of freight products, and over 12,000,000 pounds of milk in addition to furniture and loads of posts, lumber, and potatoes on return shipments from northern Minnesota. Drache found that it was easier to employ drivers for freight and milk than for livestock. In 1943, the livestock business dropped sharply and again in 1944, while freight increased to 10,907,631 pounds of which over 6,000,000 pounds was from OCCO. Milk increased even more. In 1943, he purchased his first semi trailer rig, which was devoted almost entirely to handling canned goods and beer. Several single truck operators such as Walter Klamm took over the livestock business.

Klamm became the trucker manager for the remnants of the Meriden Livestock Shipping Association associated with Central Cooperative in So. St. Paul. The various county livestock associations closed their yards and merged into the Steele County Livestock Shipping Association. Herman Larson, Lester Buecksler, Lyle Beese, Kenneth Brase, and Dennis Priebe were the truckers, respectively, who did the work far more efficiently than the local yards and the railroad.

Agriculture Flexes Its Muscles

From 1918 through 1929, Max Drache had always milked the same number of cows. During that period his milk checks ranged from $963 to $1,447 but in 1930 fell to $811; to $746 in 1931; and to $794 in 1932. He was penniless when he came to this country, and now at age sixty-one he had two farms paid for and purchased a new Hudson automobile, but for some reason he stopped keeping records. I know that he received some help from his in-laws, but there are no records showing how much. He was a good manager and a hard worker. He and my grandmother were hard workers and expected as much of their offspring. I spent a quite a bit time with them and I never heard him express any concern about how things were going for him personally. If he complained about economic conditions in the nation, he always blamed the Bolsheviks and Wall Street, but that was all he ever said. He knew agriculture was changing and understood what was taking place.

In January 1930, Dean Coffey, of the agricultural college, was quoted: "Much will be heard about [industrial] agriculture during the next few years, but I am firmly convinced that it will never supplant the present family-sized farm....It will prove to be the most efficient agricultural unit of the future." In April that year the Saco News column in the *Journal* reported that the Four Corners Farm Bureau unit (the southeast corner of the township) meeting featured a debate, "Resolved that children in the country have a better living than those in the cities." The affirmative failed to prove that the country children have the advantage. E. B. Clough, the OHS agricultural instructor, then gave a presentation and pointed out the advantages of each.

Author's Note: His record book was about the size of a standard checkbook and it was made of leather which he carried from Germany. The only other data it contained was about the help he gave to daughter Mollie and son Otto when they started farming. I found this in my Dad's papers.

In May, Minnesota Governor Christianson commented that a continued farm depression would injure other businesses. "Not many years ago... when the family was the principle economic unit there was no problem of preserving the industrial balance when the frontier farm supplied all that the family ate, wore, and used. The question of distribution did not worry the farmer....There was no trouble then about weights and grades, no charges of market manipulations, no worry about the exactions of the middlemen, no questioning of profits of pork packers." The same issue of the paper carried a notice that Otto E. Schueller, whose farm land was in Meriden and in Deerfield, had his land repossessed by the First National Bank in Owatonna. This was a rare event in the township.

In the 1920s, there was a national movement of youth to the cities because of the lack of opportunities in agriculture. While that was taking place the business people of Owatonna were working to avert the full impact of the national depression. The Gavel Club of the city, which worked to unite the local firms, was leading the campaign. Dr. F. V. Betlach, president of the group, stated, "The firms have provided means for the general depression to pass Owatonna almost completely and to prevent unemployment...from affecting Owatonna to any material degree...while noting Steele County's diversified agriculture along with the diversified businesses had been largely responsible for the communities' progress." Bank president Paul Evans "lauded Steele County agriculture for establishing an income more than $5,000,000, half of it from butter, approximately $3,500,000 of which flowed through Owatonna....But the six firms cited employ 650 persons with a total payroll of $850,000 and a business volume of $6,500,000.... These industries all started from scratch and it is interesting that men rather than money is accountable for their success." The men cited were C.I. Buxton, Thomas Cashman, George Anderson, O. H. and A. M. Josten, Dan Gainey, R. A. Kaplan, L. C. Lange, Henry Hartle, the Hammel Brothers, Joseph Kovar, and Fred Johnson.

Contrary to the national movement cited above, the number of farms in the county increased in the 1920s from 1,860 to 1,925 in 1930, and at the same time the number of cows decreased by 10 percent, but production was up. Horses decreased by 11 percent, a sign that farmers were mechanizing and thereby increasing production at less cost per unit. Test plot reports for 1930 indicated as much. Barley yielded 52 bushels an acre, Gopher oats 99, Kherson oats 107, and Minnesota 13 corn 71.8 bushel, all new highs.

At the same time the college of agriculture was warning farmers that "horses affected by over heating are never the same. Work them early in the morning and in the evening rather that during the heat of the day. Give them an extra pail or two of water in each period and turn them into the pasture at night instead of a hot stuffy barn."

In October 1931, Farm Bureau sponsored the county corn husking contest on a farm near Ellendale, which 500 attended to watch Lloyd Stutesman, of Berlin Township, compete with ten others and win with husking 1,189 pounds of marketable corn (that is, after the husks are removed from the ears) in eighty minutes.

Because 1931 was a dry year, most of the farmers were nearly out of hay by April 1932 and were waiting for warm weather to push pasture growth. That spring two carloads of beet pulp were shipped to the village to supplement the depleted hay supply.

The town board decided it was time to do something about weed control and adopted the Redwood County Cooperative weed control plan. The most mentioned weeds were leafy spurge, Austrian field cress, sow thistle, Canadian thistle, dodder, creeping jenny, and French weed. French weed, often referred to as stink weed, when eaten by cows, gave a very undesirable taste to dairy products, but leafy spurge was pointed out as being the most dangerous weed. One farmer in every section was designated as section chief. The hoe and the scythe were the chief weapons in attempting to control those weeds, but as a last resort hands covered with thick gloves were used. Being assigned to control weeds probably caused more hired men to quit than any other job on the farm.

The following data of thirty-nine farmers in six counties adjacent to Steele, although there was no one from Steele in the program, gives a good idea of what happened in agriculture between 1928 and 1931:

Date	Gross Income	Operator Earnings
1928	$5,549	$1,564
1929	$6,105	$2,329
1930	$5,689	$583
1931	$4,049	$503

This was after a 5 percent return on the farm inventory.

John Tuerk and John Karaus were optimistic and both purchased tractors. Did they suspect or did they know that tractors were more efficient than horses? Fred Mundt told the Meriden correspondent that labor and material were at the lowest point in his memory, so he decided to build a 134 x 50 x 40 foot structure of tile and lumber with a concrete floor and all-steel equipment. Hayes Lucas furnished everything for what was then the largest barn in the county. Carl Wilwock did the concrete work, and William Ahlers, a township native, did the carpenter work.

In June 1935, H. S. Gilkey, a Minneapolis lumberman, expressed his feelings about agriculture when he purchased a 275-acre farm in the township. At the same time he purchased the outstanding Winship-Nielson herd and established one of the leading registered dairy farms in the state.

The extension service was at a loss about teaching better ways of farm management because many farmers had virtually no data about their costs. More information was needed, so the extension service continued to plead for farm record-keeping cooperators so they could better determine good practices.

A Rural Youth study showed that "large farms had advantages because of the use of power machinery and on the large-scale operation the overhead is less per acre—if the farm is under good management....The size of farm which can be operated...is limited by the capacity to supervise."

The not-so-depressing news was that the county corn husking contest would be held at the D. R. Lindesmith farm near Clinton Falls.

No reason was given, but the January 17, 1938, issue of the *Journal-Chronicle*, was its last. The major national news in that issue was that the "Country is in Recession—Uncertainty....The recent tendency to assail business and blame it for the recession has not proved conducive to the cooperative spirit so essential to confidence and recovery."

The first issue of the *Photo News* on June 30, 1938, reported on a new plight for the farmers. In 1937, grasshoppers had hit the county and damaged 40,200 acres of land. To fight the pest another five carloads of "poisonous" bait were shipped in to supplement the 100 tons of bait already on hand. Farmers were reminded that they should spread the bait as soon as grasshoppers appeared. Henry Beese was the Deerfield chair, Jay Spinler in Lemond, and Erwin Wilker in Meriden.

A bit of irony in the October 1938 *News* reported that a carload of butter was consigned to Owatonna for relief purposes. The shipment came from "dairy centers less endowed than the nations' richest dairy farming section," was all that was needed to be said. In 1938, twenty-five corn test plots were planted with hybrids in Clinton Falls and Summit townships, which yielded an average of sixty-four bushels per acre with a range of fifty to eighty bushels. Only three years later Leslie Gasner became the state 4-H Corn Club champion with a yield of 131 bushels per acre. The plot received ten tons of manure and was fall plowed and planted in thirty-six-inch rows with twelve-inch spacing. It was harrowed once after planting and then cultivated four times. The increased yields came none too soon, for with the war news and increased demand for food, the corn allotment for 1942 was increased 10 percent.

In the previous chapter threshing days were discussed. A 1939 picture of threshing on the Emil Heinz farm showed twenty-three people working on the actual threshing and three adults plus two children serving lunch. They had threshed 1,201 bushels of an oats/barley mix from seventeen acres. But by 1928, two men with a pull-type combine and two people to drive grain trucks with far less physical stress had harvested more grain than the entire threshing crew pictured. Mechanization was having its impact on farming, for the 1940 census indicated that the county population had grown to 19,749. All but 234 of the increase took place in Owatonna, and the rest was in Blooming Prairie, Ellendale, and Medford.

A classic example of what could be done in farming during the period covered in this chapter is exemplified in the career of two Meriden natives I knew. A woman started working as a hired girl in 1926 when she was thirteen and earned $5 a month plus room and board. When the economy declined, her pay was reduced to $3 per month. In 1930, she was seventeen and had another job as a hired girl taking care of the Grandprey twins and earned $4 per month plus room and board, but that job was easier than working for a farm family. During the same period the man worked as a hired man for Blaine Tuthill, where he received $25 a month plus room and board. The couple married in 1932 and rented 160 acres and received some help from the man's father. In a couple of years they purchased the farm they were renting for $6,400 and had good luck with clover seed, which helped them pay for that farm. In 1937, they purchased eighty acres for $5,000 and the Hope banker, Ted Kuchenbecker, loaned them what they needed. Then they rented another 120 acres. In 1940, he had to take a physical for the service, but they had four children and he was exempted from the draft. By then they had a WD Allis Chalmers and a six-foot Case combine. "In 1944 we paid off all our debts. We had our corn cribs full and the price went to about $1.70 a bushel so we sold the entire crop and soon after the price dropped." They retired from farming in 1959, built a house in Owatonna, rented out their farm, and continued working until he was sixty-seven. He was ninety-seven and she ninety-five when I interviewed them, and they appeared to be in good health.

Specialty Crops

Railroads started campaigns for better farming methods as soon as they had their lines completed. They worked with extension people to help the cause because it was to their interest but also to the benefit of the farmers, the farm communities, and the rapidly growing urban industrial society. The Owatonna Gavel Club members understood what thriving agriculture meant to the community in so many ways. In March 1930, the Club, with the aid of local service clubs, the Farm Bureau, local cooperatives, and the CNWR, together with the extension service, provided a "Seed Special" of seven cars, four of which contained seed exhibits. Despite unfavorable weather over 400 farmers gathered in Owatonna to hear talks and observe new seed samples. A railroad officer expressed their interest as follows: "It is a well known fact of our history that our abundant agriculture was a prime mover in our society and it was widely understood that the farmer who could produce for the lowest cost will take the market and survive." Bob Hodgson, superintendent of the Waseca Experiment Station, stressed that improved corn varieties would produce 12 to 15 percent more per acre, but big losses still came because of poor weed control. He said, "From 1900 to 1921 an average of 3,963 car loads of dockage were taken out of the grain at the Minneapolis mills each year."

As stated previously, sugar beets were introduced to the township in 1928 and were successfully grown and shipped. In April 1930, the Surge Company announced that acreage would increase to 392 acres and allotted them to eighteen area growers. William Behsman had the largest allotment of sixty-five acres, followed by H. W. Schuldt with forty, Gus Fette and L. P. Zimmerman with thirty each, and Paul Krenke with the smallest allotment of four acres.

In 1931, the allotment was reduced to 300 acres because the warehouses still contained much unsold sugar. That fall Otto Schueller invented a beet loader, which proved successful and made loading easier and faster in his operation. That year sixty-five cars of beets were shipped from Meriden. Delbert Brase and Robert Schuldt did the weighing at the local scale. A few acres were not yet harvested.

In April 1932, the American Beet Sugar Company announced that no sugar beets would be taken from the Meriden station that year, and in May 1933 the beet loader and scale were removed from the village. In 1932, it allotted seventy acres to seven growers, but all beets would be delivered to Owatonna. In the spring of 1933, three growers traveled to Mason City to arrange for a beet contract but they had to deliver to Owatonna or Waseca. That appears to have been the last year for beets. The only yield figure that was found in the paper was that Oliver Stendel got fourteen ton per acre on his first year and ten ton in his second, which would have been competitive to corn at that time except for the labor required. Difficulty of getting labor necessary to raise beets and improved corn yields made growing corn more attractive.

The depot, the beet hoist that raised the beets dumped into a hopper pit from trucks or wagons and then lifted to the rail car, the scale house where the trucks and wagons were weighed. That is the chief purpose of the photo in this book, but it was taken to show the 1932 "Meriden gang." L-R: Eleanor Arndt, Jeanette Palas, Dorothy Arndt, Barbara Arndt, Laurel Drache, Hiram Drache.

In 1936, L. C. Palas shredded corn at the John Reudy farm north of Goose Lake and reported that the hybrid purchased from a seed salesman from Waseca yielded ninety bushels per acre. The second reference to hybrid corn was that Neil Young, who graduated from OHS in 1937, was the first in the county to plant hybrid corn in his plots for his agricultural project.

Canning crops had started to make inroads in the 1920s, and the total acreage increased ten fold by 1930. By then the county ranked high in the state in sweet corn and pea production. That year OCCO contracted for over 500 acres within six miles of Owatonna. Two pea viners were west of the city and took the crop grown in the township. The company produced 48,000 cases of peas in 1930.

In 1932, a Drache truck was contracted to haul peas from the farm to the viner and haul sweet corn directly to the factory. The windrowed vines were elevated onto the trucks, which were unloaded at the viner and then were hand forked into the viner. The peas were hauled to the cannery within hours after the crop was cut, and the pods and vines were stacked by the viner and sold for cattle feed. The ensiled vines emitted a distinct pungent aroma as they were hauled away from the viner. Both corn and peas were successful crops, and by 1933 the state was fourth in the nation in pea production and first in sweet corn.

The Henry Scholljegerdes corn shredding rig c.a. 1936; the shredding was blown into the hay loft and used as feed and bedding and the ear was shelled for hog feed or ground for the cattle. Robert Scholljegerdes photo.

After World War II started, OCCO acquired the former Kovar Manufacturing building and established a dehydrating plant. By 1943, that plant processed 1.5 million pounds of potatoes, carrots, and beets annually. The entire production was contracted by the U.S. Government. It operated from November until May on a twenty-four hour basis with three shifts of fifty employees. The crops were all grown in the Hollandale area. In 1944, 50 percent of the production of the cannery was requisitioned for the military.

Diversified Farms

Sod breaking was a particularly laborious and time-consuming process. Unless farmers had the resources to hire the work done or purchase the horses and hire the help to do that work, only a few acres could be broken each year. After they had more land broken they were able to support cows, chickens, and hogs, in that order. They usually started with limited numbers of each so they could provide a better diet for their family.

Chickens were often taken on the wagons as families moved west because they were relatively easy to transport. Many times a cow or two was tied to the wagon or the children herded them behind the wagon. After production was beyond the family needs, the eggs traditionally were traded for groceries. Stanley Newhall, from Clinton Falls, was a member of the Better Farming Club and was one of thirty farmers in that organization who kept records on his poultry enterprise. In August 1930, records of his flock indicated that they averaged 192 eggs per year. The club average was 108, and the lowest flock yielded only 50. Newhall earned a return per hundred hens of $518.29 and his net over feed cost was $367.76. In September 1938, John Tuthill reported that he had a large flock of New Hampshire Red chickens. Farmers seemed to prefer the heavier chickens because they were as interested in having chicken dinner as they were in having eggs.

Initially, farmers hatched their eggs by letting clucks sit on the nest ,but as the flocks increased incubators were used. My paternal grandparents had a kerosene incubator that was brought into the parlor, the most remote and least used room of the house, each spring and the eggs were hatched there. My wife said her folks had an incubator in the entry way to their house. As soon as the chicks arrived they were taken out to the brooder house, which was kept warm by a kerosene stove. Other than the peeping of the chickens, I never detected anything out of the ordinary when hatching was taking place.

As more farmers developed larger flocks, commercial hatcheries appeared. In 1940, Collins Produce, in Owatonna, added a new 31,360-egg incubator, which doubled their capacity to 62,720 eggs. In 1941, the Snow Hatchery opened a plant in the vacant Ganser brewery on South Oak Street. This was their twenty-ninth location, an indication that the poultry industry had become commercialized. A survey at that time showed that 55.7 percent of the farmers were keeping lights on in their poultry barns.

In August 1930, Ben Kuckenbecker had a net return of $4,321—the highest in the Better Farming Club. He credited his three-year rotation of corn, oats, or barley seeded to red clover for one year. All the grain, hay, and straw were used by the fifty milk cows, young stock, and horses. Up to that time he had not applied any commercial fertilizer. There were 105 dairy herds in the county that were in the Minnesota Mail-in Cow Testing Association. In 1931, J. C. Penny, the merchant, had a fourteen-year-old cow in a herd he owned that produced 17,235 pounds of milk and 959 pounds of butterfat, which was a world record. The next year Femco Farms at Breckenridge, owned by F. E. Murphy, a Minneapolis publisher, had a cow that set a new record of 35,626 pounds of milk and 1,483 pounds of butter fat. The news article read, "Her equal may never be seen again."

In 1930, the Andrews family had the longest continually owned farm in the township. James W. Andrews died in 1919 and his three sons, Harry, James, and Clayton (Bud), formed Andrews Brothers and continued a diversified operation until about 1950. The farm was over 300 acres, and they had the usual chicken and dairy operation, but they realized that they had to expand one of those enterprises to gain economy of scale. They built a modern dairy barn and increased the dairy herd to forty milking cows. A second barn was built for eight draft horses, a bull, and all the young stock. During this period Harry was active in the Meriden creamery, one of the twenty-two cooperative and one private creamery in the county that produced 5,384,414 pounds of butter. This was 63 percent more than had been produced a decade earlier. The county claimed to be "The Butter Capital of the World" based on butter produced per square mile. John Tuthill followed his neighbors, the Andrews, and in 1934 built a 32 x 84 foot barn with a 20 x 30 foot "L" addition with all steel barn equipment.

The horse herd at the John Tuthill farm in 1934 taken to promote his new barn.
Tuthill raised horses and was quite outspoken about their advantage over tractors.

In 1936, after a county-wide tuberculosis testing, thirty-six reactors and twenty-five suspect cows were found and disposed of, which accredited the county for another three-year period. At that time the county had 498 herds with 13,701 cows for a herd average of 27.51. The township had no reactors. Reuben Mundt, who had the largest herd in the township, was probably the first to own two double-unit milking machines. The dairy industry tried to stop the sale of oleo margarine in the 1930s, but after several years of campaigning they realized they could not win and gave up.

On March 1, 1940, the Meriden Dairy Herd Improvement Association was organized by twenty-three farmers under the leadership of W. P. Jones, who owned a dairy farm, and Roy Bakehouse. Dale Jones was the supervisor and Leonard Flatten the cow tester. Within a few months the association had forty-seven members. The first years' records indicated that the ten low cows produced 161 pounds of butter fat and brought a return of $15 above feed cost. The average for all 606 cows was 295 pounds of butter fat for only $7 more feed and earned a return of $45 above feed cost.

In 1942, most of the members of the association signed up for the artificial breeding program because it was the quickest way to increase the quality of the herd, plus it eliminated an annual feed cost of $75 for the bull and the risk of having a dangerous animal around. In 1943, the Southern Minnesota Breeders Association completed its first year of artificially breeding nearly 5,000 cows. Of the 15 sires, seven were Holstein, four Guernsey, two Jersey, and two Milking Shorthorns.

By contrast to what members of the association produced, the Gilkey Farms had an aged cow that was milked three times daily and produced 25,877 pounds of milk and 890.5 pounds of butterfat. The average for the entire Gilkey herd was 520 pounds of butter fat, and for those being officially tested the average was 730.8 pounds.

In 1941, when butter maker Loren Luhman resigned, Arthur Uecker, who had been employed by the creamery in 1938, assumed that position. In 1941, the creamery produced 320,000 pounds of butter, and each following year the volume increased until it hit 1,657,424 pounds in 1965, which was its final year. The dairy industry entered a new phase of production, and local cooperative creameries became extinct.

The traditional saying among farmers in the pre-industrial era was that the chickens provided the grocery money, the cows provided the cash flow to keep the family and farm going, and the hogs were the mortgage lifters. In 1930, Louis Radke, one of the early Better Farming Club members, said that his 154-acre farm operation included nineteen litters that produced an average of eight and one-half pigs, which yielded a hundred weight of pork at a cost of $7.78. That year his hogs yielded a net return of $5.54 over feed cost per hundredweight. Radke also had a leading dairy herd.

In 1939, over 500 farmers attended a hog cholera school sponsored by the extension service. I remember the fear in the area that year when many farmers in the township lost hogs. In 1941, Roy Bakehouse, one of the first with a college degree who farmed in the township, was elected president of the Owatonna Swine Improvement Association. Melvin Froelich, also of Meriden, was secretary/treasurer. The goal of the association was to check litter weights at fifty-six days and then at sale time so they could select the fastest growing strains and then keep improving the feeding and management practices. Tom Raine, the OHS agricultural instructor, was their leader. That year extension agents informed hog farmers that self feeding was recommended because "it was not only a labor saver…but pigs will make much faster gains and stand a better chance of yielding a profit…. Pigs are especially adapted to self feeding….They will eat at all times of the day or night….There is almost no danger of over eating or digestive troubles."

The above are some of the breakthroughs that took place in the township and nation during the 1930s and early 1940s. Agriculture would move even faster in the years ahead as the historical truism that the very best is three times the average was proved many times over.

A New Direction for Agriculture

As stated earlier the production of crops and other agricultural commodities increased steadily during the 1920s, and much of the most favorable tillable land had been put to use. Farmers were mechanizing rapidly and tractors had taken away the need for horses, which freed millions of acres for producing other crops. Farm labor became more efficient, which also freed many farm youth, who looked to the city for a better future. The government reacted to the problem, and in 1929 the Federal Farm Board was established. Its declared purpose was "to place agriculture on a basis of economic equality with other industries." The basic idea was to unite the farmers into cooperatives and control their production. In 1930, the Grain Stabilization Program was enacted with the power to use federal funds at steeply discounted rates to purchase surplus production and withhold it from the market. Only the Federal Reserve had greater power. Owatonna native Congressman Victor Christgau saw what was happening, and in 1931 said that "cooperative effort to reduce production is insufficient, crop control need is evident."

Conditions in agriculture were desperate, but it was much worse in many other sections of the nation than in southern Minnesota. In August 1932, the State Farmers Holiday Movement expected to have 2,000 to 3,000 turn out for a meeting, but the local papers reported to the national leaders, "Farm strike is not receiving any marked degree of attention in Steele County to date, other than occasional comment." In September, Thomas Cashman was selected the temporary chair to lead the Holiday Association and was assisted by with five others from Deerfield, Lemond, Meriden, and Owatonna townships who acted as area leaders. Later that month, Milo Reno, the legal mind of the Holiday Association, spoke to a crowd in the court rooms and "hit at Hoover and the Wall Street gang—and said that some townships had 75-85 percent signup. He said his talk would be nonpartisan and nonpolitical." In October, after a plea from Thomas Cashman, the Owatonna Association, which represented the mercantile group of the city, voted: "Our moral support and assistance towards its success." About 200 gathered to urge that Steele County farmers should join the national "farm strike," which claimed that 85-90 percent of the farmers had signed up. However, in the following week the farm strike was ended by the Holiday Association, and Cashman "expressed his belief that Steele County could not gain by continuing restriction of sales."

In March 1933, extension economist S. B. Cleland stated: "The outlook for immediate agricultural relief was poor as the result of a highly involved agricultural situation. Farmers…must become reconciled to a long and gradual period of readjustment." The following week, O. B. Jesness, also from extension, commented that farmers must get export outlets. "The dream of self sufficiency loses sight of the fact that we require some products that we do not produce…for instance rubber, coffee, and tea. We produce some of our requirements of sugar, wool, and hides but not all. The farm industry will continue to be interested in export markets and we might as well develop plans on that basis."

In May 1933, the Agricultural Adjustment Act (AAA) was passed to give farmers relief via devaluing the dollar, which was intended to raise the level of commodity prices to prewar levels through a voluntary crop reduction program. Wheat acreage in the county was immediately reduced by 20 percent. The corn-hog program called for a 20 percent reduction in corn acreage and a 25 percent reduction in hog production. I remember this very well. Farmers who had engaged my dad to do their livestock hauling came to him for evidence of what their hog shipments had been for the period required by the government. Dad had records for every farmer who called on him for help. I was assigned to leaf through all the check stubs and the receipts that the trucker had from the commission firms and/or packers. I saw how desperate they were for the information and I was amazed at their total lack of records. Those who could not prove what they had produced were paid on the basis of average production for each county. Those who could prove their production, which was often higher than the county average, profited through their-record keeping effort.

In 1933, over 400 farmers turned out for a meeting when Cleland, of the extension service, gave an address about the new federal land bank system, which was offering thirty-six year loans at 1 percent. This was about the same time that the government announced the devaluing of the dollar in a effort to boost prices. In 1934, the total county income from the corn-hog program was $256,000, which was paid to 1,144 farmers in the county, 80 of whom were from the township. Their operations included 778 acres of land and 6,628 hogs in the program.

One newspaper editor wrote: "We don't bet on politics but we are willing to haphazard one little guess: we predict that the corn-hog bonus checks will arrive before November 6 [election day]." The national Farm Bureau president was more positive. He felt that the farm program had not only saved the farmer "but was the saving factor in preventing a major economic disaster."

In March 1937, the Owatonna Rotary invited Frank Peck, who had resigned from the Farm Credit Administration, to speak to them and local farmers. Peck said the AAA and Soil Conservation were sound, "but their administration has tended to subsidize more freely the incompetent farmer unwilling or unable to attempt a solution for himself." This was the reason he resigned. He continued, "Rehabilitation of the underdog and marginal producers is essential and should be done but not with an emphasis on inefficiency as is the present trend. One-third of the farmers in America operate on marginal farms. The farmer should have the right of self improvement, free intelligent thinking, and to direct his own affairs. Agriculture must not be hamstrung by a maze of government regulations."

In June 1937, Steele County was among the first twenty counties in the state to have farms mapped by aerial photography in connection with the farm programs. Other news was that the government expected a deficit of a half billion in a total budget of $9.6 billion. Five months later the deficit was a billion dollars greater and "the $150 million in corn relief loans may not be forthcoming."

Farm to Market Roads

In February 1930, it was announced that in the summer Highway No. 7 between Owatonna and Waseca would be paved. The rail siding in the village was selected to receive materials for the paving job, which enabled the construction crew to operate the shortest distance to where the actual paving was taking place. This meant that the paving crew had to be accommodated in the village. The Paul Draches were engaged to provide three meals a day for twenty-two to twenty-four men and sleeping facilities for twenty-four in their home and in two houses they rented on "the east end." The other twenty members of the crew slept on cots in the vacant church and ate their breakfast and evening meal in the little house east of the Fette garage. Drache also provided a noon meal for those twenty. As stated previously, Herb Fette had closed the garage to take a job elsewhere, and the construction company rented the garage and their house. Paul Drache used his 1926 Dodge sedan to carry the noon meal to the paving site. Those who worked at the rail unloading site, or in the garage, or drove , ate in the Fette house. A couple of times I was allowed to ride with Dad, and I can still smell the aroma of scalloped potatoes—a real favorite with the crew. The folks had a hired girl to help with the cooking, taking care of the bedding and laundry, and even milking the cows. The post office was Mother's first obligation, but she was able to help much of the time. The head of the construction crew had his bed and desk in the north porch adjacent to the entry of the post office. The Draches received a dollar a day for meals and twenty-five cents per bed.

Rosina Wilker getting the cows from the vacant lot east of the Hayes Lucas store where they had been pastured for the day. She also carried water to them at midday. Besides helping in the house she also milked the cows and sometimes delivered the milk if others were busy. Her pay in 1931 was $3 per week plus room and board, which eventually rose to $6. Rosina Schultz Kopisckhe photo.

Thornton Bros., of St. Paul, were the low bidders on the eleven-mile stretch of road nearest to the Meriden site at $240,102. The equipment was unloaded in Meriden and paving started on May 4 north of Goose Lake and worked east toward the Meriden junction. When that section was finished they started at Crane Creek school corner and worked west to the Meriden corner, which was finished the last week in August. After the paving was completed, a crew built the shoulders and erected signage while the paving crew went to Owatonna and worked west and then to Waseca and worked east. The shouldering for the road between the two towns involved moving 82,000 cubic yards of material at a cost of $40,103. When the highway was finished, Owatonna had four concrete radiating roads and had the distinction of being the smallest city in Minnesota to have three trunk highways. In January 1934, Highway No.7 was renumbered U.S. 14 and probably will be relegated to a county road when the current (2012) four-lane route across southern Minnesota is completed.

The income from those 120 days of catering to the highway workers was a real boon for the Draches. They invested $1,000 in a piped air furnace with registers to all the rooms in the house except the tiny kitchen, which was heated by the cast iron wood-burning stove. They also installed a chemical toilet in the basement, near the furnace. Real toilet paper was supplied so no more catalogs or peach wrappings were needed! Mother got a new square-tub Maytag washer with ringers, and Dad had $550 for a new truck and concentrated on trucking.

The nation's love affair with the automobile seemed to know no limits. In 1921, Ford Motor Company surpassed one million cars for the first time and maintained that volume each year up to 1929 and again in 1930 and 1935. In 1936, 1,219,262 cars were sold. By 1936, there were 24,197,685 cars and 4,023,606 trucks registered in the nation. In August 1931, the daily traffic count on Highway 14 through Owatonna was 1,375 cars, which was the greatest increase of any road out of Owatonna for the first year of pavement. The daily count out of Owatonna to Faribault was 2,897, and to Austin 1,086.

The paper reported that highway construction was lagging behind demand, but the good news was that the cost of maintaining unimproved state roads was $1,458 per mile while on paved roads it was only $279. In 1937, Steele County reported that the average cost of maintaining a mile of county road was $671 while the paved county roads varied from $304 to $484. If it had not been for heavy snow it would have been even more favorable for the cement roads. But the cost of using the highways rose as traffic increased and society decided that people needed to be protected from themselves. More patrol officers were needed to control speeding and to test automobiles for lights and brakes.

Muddy, impassible roads became less of an obstacle to driving as county roads improved, but in March 1937, a snow storm passed through from the village to the highway corner and the road was blocked from Saturday to Monday afternoon. In some areas the snow was so deep that horses could not travel, and shovel crews had to work ahead of the county plows. By 1937, the county owned 124,700 feet of snow fence to help against drifting in the twenty-four miles where the greatest trouble occurred of the 200 miles of maintained county roads. That year the county purchased a five-six ton fully equipped truck plow for $11,840.

A snow storm in the spring of 1936 left the snow packed so hard on roads that the county road supervisor called and asked that the banks be broken to enable the plows to get through. Three WWI four-wheel-drive army trucks in tandem pushed the plow. They entered Meriden from the south and had to ram the bank that the men had shoveled on before the road was opened. I was able to identify all but two. Front row one, 1. Hiram Drache with helmet, 2. Lloyd Grandprey with his dog Fussy; row two, 3. Harvey Moore white shirt, 4. John Abbe, ?, 5. Donald Born with helmet, 6. Walter Klamm with shovel handle, 7. Paul Klamm, 8. Dallas O'Neil; row three, 9. Benny Anderson, ?, 10. Henry Wicklow, bow tie, 11. Albert Born, 12. Art Willert with tassel on cap and hands folded; back row, 13. Loren Luhman, 14. Martin Fette, ear lappers, 15. John Evans, 16. Ray Evans, 17. Herb Fette.

Traffic kept increasing as school buses were on the roads daily, some farmers or their wives had jobs off the farm, and milk trucks were on their daily pick up. Farmers and their wives often by-passed Meriden and went to Owatonna or Waseca to shop. The route for the mail carrier was enlarged because everyone assumed that the roads would be open.

Daily Life and Culture

Everyday life changed because of the rapid improvements in electrification, roads, automobiles, mechanization in agriculture; industrialization; and then by the demands of World War II. A perusal of the St. Paul's Lutheran Church records indicates that up to the 1930s most of the marriages were between members of the congregation. After more members started going to high school, many of them married people from other areas. Some veterans married people from other states and foreign countries.

I started high school in 1938, and prior to that very few from the congregation had gone beyond grade school. Neither of my parents went to high school, but Dad had a year at Canfield and a two-year Commercial

Course at Wartburg College. Mother was there one year. This is an isolated case, but I am aware of a marriage that took place in another community in 1952 where tears were shed because someone from the congregation married someone from another congregation even though it was in the same denomination.

The Worthwhile Club, in the eastern part of Meriden, southern Deerfield, and western Owatonna townships, held their July 1930 meeting at the Albert Nielsen home. The program consisted of a stereopticon lecture on "The Home of Evangeline" by Rev. Ilsey, followed by accordion and Hawaiian music by local performers. At its February 1931 meeting of the club, former members were invited to honor Mrs. H. J. Bundy, who had moved to Iowa. Cootie was played at six tables, while many others preferred "just to visit." A Valentine party was held at the F. W. Schuldt home with forty-five guests playing cards at seven tables. At the same evening and only a few miles away another group played hasenpfeffer at the Arthur Heinz home.

In August 1931, the Meriden Village Club was organized for the year and had Dr. C. L. Melby, an Owatonna chiropractor, speak on how to maintain good health. The club's project for the year focused on nutrition. At the next meeting the lesson was on Food Needs and Food Selections. Daily food records of each of the members were examined. Mrs. K. A. Gontarek won the whistling contest, after which refreshments were served. The nutrition project was continued by the club for four years. The final year dealt with nutrition for the growing child and exercises to maintain proper posture.

A large gathering met at the hall to honor Mr. and Mrs. Rudolph Schendel on their 25th anniversary. Dancing provided the entertainment after which lunch was served and a purse of silver was presented. The Worthwhile Club had a debate, "Resolved: that it is justifiable for married women to 'Do Work Out-Side the Home.' The negative team was declared the winners. The Business and Professional Women of Owatonna were opposed to the legislature's attempt to pass a law limiting women to working not more than forty hours per week because that would cause discrimination against them in businesses where men were allowed to work longer.

In April 1934, the roll call for the Worthwhile Club was answered by garden hints, and the program consisted of a seed and plant exchange. Martha Ratte, an extension soap specialist, conducted a soap making demonstration at the hall. The soap made at the demonstration was given to the woman holding the lucky number. In 1935, nearly 500 women participated in the homemaker's project "Problems of Home Management." Mrs. Lloyd Grandprey was the Meriden chair of the event. The Meriden Boosters Club and the Meriden Village Club both conducted "Home Betterment" projects that year.

In 1936, over 500 rural women were present at the extension program in Owatonna. Mrs. Reuben Pieper, Mrs. Henry Palas, Miss Leona Schuster, and Miss Gertrude Abbe represented the township clubs that year. In 1937, both clubs had projects on home furnishing, refinishing furniture, caning chairs, removing old finishing, and refinishing, after which all clubs of the county attended a pageant in Owatonna where the best examples from each club were displayed. In October 1940, the biggest cooking school to date took place when 800 women swarmed the armory for two days to hear national home economists discuss preparing new foods and how to use all appliances.

The western part of the township apparently did not have a 4-H Club until some of the women in the village led by Mrs. W. P. Jones organized what was called "The Meriden Hot Shots." I was a member of that club and my project was sewing. One of the things I made was a shoe bag that contained pockets for four pair of shoes and was hung on the inside of the closet door. In the second year I had a dairy calf and got to spend three days at the fair with others who also showed an animal. After the Jones family moved to Owatonna the club disbanded, and it was not until 1939 that Lloyd Ebeling and Mrs. Reuben Ebeling established a club. Now Steele County could boast that it had a club in every township.

R. E. A. to the Rescue

In the previous chapter it was stated that in 1923 Howard Deichen had introduced electric power into the village and to the farmers who lived along his lines and were willing to pay the $500 connecting fee. Those who lived in the eastern part of the township were able to connect to the lines that ran into Owatonna in many directions. Each year more farmers who lived along the existing lines learned the advantages of the service. But many farmers did not live along power lines and wanted the service, so in January 1936 about 600 farmers met in Owatonna to organize a rural electric association. Other meetings were held throughout Steele and Waseca counties, and on October 10, 1936, over 800 farmers voted to merge their efforts into the Steele-Waseca Cooperative Electric Association. In November a contract was signed to purchase wholesale power at $0.0277 per kilowatt.

In June 1937, poles began to arrive in both counties. In July L. P. Zimmerman, a Woodville Township farmer with a degree in electrical engineering, was named the project director for erecting the power lines. On July 5, Senator George Norris, "the father of rural electrification," spoke at the pole setting at Beaver Lake to "set off one of the largest electrical co-operatives in the state with 310 miles of lines scheduled." Herb Fette was quick to catch on how using electricity could help the farmer for, in April 1937 he had developed and already was selling an electric fence for $18.95. The electric fence was not only a real labor saver but it enabled dairy farmers to rotate their summer pastures, which proved to be a boon to both beef and milk production.

In October 1941, a greatly enlarged cooperative moved its offices to Owatonna where it could better oversee its 900 miles of lines and 1,828 members. A survey at that time showed that 94.4 percent of its household patrons had an electric iron, 75.5 percent an electric radio, 95 percent an electric washing machine, and 47.7 percent an electric toaster. Of the farm uses: cream separators, 77.3 percent; poultry house lighting, 55.7 percent; milking machine, 34 percent; and electric fence; 17.6 percent. District No. 26 in Havana Township was the first rural school in the county to have electric heated hot lunches. That was an improvement over putting potatoes, hot dishes, or cocoa on the space heater.

Because of its location, Meriden had a greater percentage of its farmers on lines of the public power, but rural electrification transformed rural America. The local movement was part of what became the Rural Electric Administration (REA) that was later transferred to the USDA. By 1942, it had more than five million rural members. That is a good indication of how rapidly electric power penetrated the rural areas.

Schools

In 1929, the state began to put more pressure on school districts by developing a separate state examination for districts that still had eight-month terms. In July 1930, the county rural class was 138, the largest number to get diplomas at the ceremonies held in the high school auditorium.

Effective September 1930, the state required that all state aid schools would have single seats only thereby negating the 1920 rule that allowed double seats. Marie Christianson, the county Superintendent of Schools, in cooperation with the state, was trying to force all rural schools to adopt a nine-month school term in order to make it easier to conduct the state exams and in hopes that it would improve the graduation rate. At that time only twenty-three districts out of seventy-nine had nine-month terms. This was in contrast to 1887 when only one rural school had a nine-month term.

In 1932, County School Day was inaugurated in which 400 rural school youth gathered with 2,000 parents and rural teachers at OHS. The reason for this was to get rural students more acquainted with OHS in an attempt to encourage more to enroll. The total number of students was not increasing, but in 1933 a new high of 143 graduated, which meant that more students were passing the state examinations rather than biding their time, as in 1926 when only forty-seven received diplomas. The others went until they were sixteen and they no longer had to attend. In 1936, OHS discontinued the normal teacher training department that which had started in 1912 and was the oldest and strongest in the state, but now the teachers' colleges where doing the training. By 1937, about 60 percent of the rural graduates entered high school, which was the highest in history. To encourage more rural students to attend high school, the state passed a law that paid the cost of traveling to high school.

September 1930, the first four grades in the village school, Miranada Johnson teacher L-R: first row, Hiram Drache, Iris Born; row two, Dorothy Arndt, Emery Brase, Norbert Abbe, Erwin Woker, Rodney Walters; row three, Vernon Runge, Mildred Abbe, Alfred Buesksler, Mary Jane Grose, Donald Born; row four, Vernon Nelson, Marvin Gontarek.

Two external events regarding schools happened about this time. First, Anna Drache, who had boarded and roomed teachers since September 1924 for $0.50 a day, decided that after teachers' salaries rose above $60 a month in 1936, the daily cost should raise to $1.00. It remained at that level until 1952 when she stopped having roomers.

Secondly, in 1939, Otto Woodrick, of Owatonna, contracted with several rural teachers to deliver them to their school each morning and return them home each evening. These teachers wanted to live in Owatonna rather than at rural homes as in the past. This probably reduced the high turnover of teachers. See the table in the Appendix to get an idea of the tenure of the teachers in the districts.

In 1940, farmers' night school was started at OHS. Meriden had a good delegation with Paul Buecksler; Frank Dinse; Roy Bakehouse; August, Martin, and Melvin Froehlich; W. P. Jones; and Ewald and Laverne Wilker in attendance. These were many of the same people who belonged to the dairy testing association. That year "The Christmas Seal Nurse," Mabel Johnson, conducted a rural school health check to "discover health defects and stimulate their early correction. Children were especially inspected for defects in vision, hearing, posture, nutrition, and general health conditions." A permanent set of records was established. In 1942 the cost of educating 3,928 students in the county averaged $94.00 per pupil.

St. Paul's Lutheran Church

After the closing of the German Methodist Evangelical Lutheran Church in the village, St. Paul's was the only congregation in the township with a church building. As stated in the Preface I have chosen not to include a history of St. Paul's because that story is recorded elsewhere, but I am mentioning a few events that had a bearing on the community. In December 1930, the St. Paul's Council made arrangements to have an English service the first Sunday of each month. I remember sitting through German services and understanding some of the sermon, but I recall that some of my Sunday school classmates were very fluent in German and sometimes even recited in that language. My grandparents always spoke German with my parents but always spoke to me in English.

The Rev. Carl Harrer had led the congregation from 1910 until he retired in June 1932. He was very fluent in English and very well liked, but there was no concerted effort to switch to English until 1930. If there had been my parents would have been in the front ranks to make the change. But people were moving into the area who did not know the language and more English was needed.

In 1932, after John Voelk became pastor, at one of the first meetings he increased the number of English services to 50 percent of the time. With his encouragement the council voted that all female members of legal age "shall have the right of voice and vote in all congregational matters." In September a Luther League unit was organized with thirty-three charter members. This was the organization in the congregation specifically for post-confirmation members.

The congregation was growing, and in 1935 two-thirds of the services were conducted in English. That year the congregation built an addition to the parochial schoolhouse, which provided the Ladies Aid with a kitchen. On August 27, 1936, lightning struck and destroyed the building that had served since 1876. In May 1937, the new building was dedicated in which all but one service was conducted in English. I recall the event and particularly remember when word was passed around that pickpockets were at work in the crowd. In 1941, German services were discontinued.

A Hodge Podge of Daily Events

The following items are miscellany that went on during this period and that might typify what was happening in small-town USA during this era. In 1930, Adeline Drache, age fourteen, had an operation on a goiter in her neck at the Waseca hospital. Her bill for one week in the hospital, including the operating room, was $35.00. In July 1931, Mrs. Paul Drache and Mrs. Floyd Wolters motored to Waseca with their sons, Hiram Drache and Rodney and Norman Wolters, to have their tonsils removed. The bill for each of the boys was $15.

Gus Schendel, who lived on a large lot south of Mill Street across from the stockyard, trapped rabbits to keep them from harming his garden and to provide food for himself and the trimmings for his small flock of chickens. He kept his garden fertile by hauling manure and the soil from where the stockyard had stood with a wheelbarrow. I often visited him at his very small house. He always smoked a pipe and his house was saturated with the smoke smell. In his early years he painted buildings, but all I ever saw him do was work in his sizeable garden and lawn.

One of the "happiest events" that took place in 1931 was when a local farmer, who also was a local bootlegger, had a picnic at his farm one warm afternoon. Before any of the guests arrived, he spiked the watermelons. The women who informed me said, "We were all laughing to beat the band after a few slices of melon." I knew many of those who were involved.

At 11:30 p.m. on February 23 in 1932, my mother was awake because she was treating me for an ear ache. She saw a figure enter the cream station and when he later came out with a bundle of furs, both of my parents gave chase. "The thief stumbled and dropped two of the furs, but kept running. They saw him throw the rest of the bundle over the fence between the pool hall and the Herb Fette residence. Several men came out of the pool hall and joined in the search but no trace of the thief was found." The next day Dad visited with Bill Arndt, the barber, and learned who had left the pool hall previously. It was the person Dad had suspected.

In April that year, many chicken coops were robbed. This happened quite regularly, especially if the farmer did not have a good watch dog. A few nights later a thief pried open the office window at Hayes Lucas and stole the usual pocket knives, razor blades, pliers, and other small items "amounting to about $50.00." The same evening someone attempted to break into the office at the depot. Agent Elmer Johnson stated that nothing was left in the office overnight. That was the fifth robbery since the fall of 1931.

In 1932, R. James and Loretta Stuart became residents of Meriden. James Stuart was an artist and rented the church that had been vacant since 1927 to use for a studio. They and their big black German shepherd lived in an apartment above Luhman store for several years. Two other families also lived in that building at the time. In 1934, Stuart received his major contract after coming to Meriden. He was to paint the "Butter Capital Kids" on the large billboards that Owatonna placed along each major highway that hailed the city as the Butter Capital. I remember the big billboard standing against the high church walls.

In 1936, Stuart purchased the former livery barn from Mike Gontarek who had purchased it from Luhman when he retired. The building became a real attraction when it was remodeled to be both living quarters and a studio on the second floor. Windows were installed in the north roof for a skylight.

In 1937, Stuart was called to New York by Standard Oil to consider a contract. He refused to stay in the city to do his work, but he was awarded the contract for twelve paintings for the Standard Oil calendars for 1938. Whenever Stuart was short of money and wanted to go on "a drunk," he came to my mother and offered to sell her one of his religious or children paintings. Over the years my parents collected a number of Stuart originals. Unfortunately, my father did not fully understand the value of what they possessed and in later years sold some of them to strangers at minimum prices. If you want to know more about Stuart, he had good coverage in the papers and also in a book by Sterling Mason.

The Chicago World's Fair took place in 1933 and a few from the township attended. Walter and Gertrude Voss went there on their honeymoon. W. P. and Grace Jones and son Ferris, Laverne and Benita Wilker, Joseph Kasper, Thomas Karaus, and Arnold and Adeline Drache were accompanied by Esther Scholljegerdes and another cousin. That year the Steele County fair was blessed by ideal fair weather and had a record closing day of 25,989, which helped set a new attendance record of 83,757.

The county voted to repeal Prohibition by a margin of 989 votes with less than half of total voters who had voted in the 1932 regular election. Somerset voted wet 126-30 and Meriden 110-37. They were the two strongest precincts in the county percentage wise to vote wet.

At their 1935 annual meeting, members of the Deerfield Rural Telephone Company passed a resolution "that will halt enlivening a dull day and learning the news popular with some farm women by limiting a single call to five minutes." Harry Andrews, president, and E. G. "Emil" Heinz, secretary/treasurer, both stated there would be strict enforcement.

Those who experienced living in that period know how frustrating it was if you wanted to make a call and people would not yield the line when they knew others were trying to use it. My parents had both the Owatonna and the Waseca phones and were often called by people on one line to make a call on the other. That way they could save $0.15 by not having to make a long distance call. Ironically, some of these were people who never did business with the truck service or even the post office.

Rubber-necking was a favorite pastime for some, which could be irritating, especially when the rubber necks had a loud clock on the wall near their phone. One day my mother got a call after she had just had a conversation with someone. The caller asked, "Who died? I didn't get to the phone quick enough to hear who it was." Another time someone came to the post office and asked to use the phone. When she was, instead of hanging the ear piece back on the hook, she set it down on the desk. After a long spell without getting any phone calls my mother thought something must be wrong but then saw why.

John Tuthill built a new barn in 1934, and before he used it for hay he held a barn dance. Some times dances were just for having a party, but many of them were held for the revenue. Tuthill built a substantial barn and probably was as interested in showing off the barn as he was in making money from the dance receipts. The Ross Gordy band from Rochester played; admission was $0.40 for gentlemen and $0.20 for ladies.

In 1934, the Hormel workers at Austin conducted a prolonged strike. I was with my dad when he delivered a load of hogs to the plant. After we left the plant some strikers tossed a log on the road in front of us. Those mechanical brakes were not able to stop "on a dime," and we both felt the bump. Later that year, Dad was delivering a load of canned goods to the Twin Cities and I was along as was my uncle Arnold. Apparently our load had some priority because when we got to Fort Snelling two National Guardsmen got on our running boards and two sat on the front fenders with their feet on the bumper. A total of 3,700 Guardsmen were called duty because Minneapolis Teamsters 544 was on strike. We saw action that day that made us appreciate the troops who were on our truck. In 1937, Tony Gontarek, of Meriden, was driving a semi trailer of milk for Albert Lea Food Products Company. Six miles south of Wells he was stopped and beaten by six labor organizers. He was "left lying on the road side and his truck was in the ditch."

Two other events took place in the 1930s that reflected a trend that was gaining momentum in rural areas. In 1933, the Clinton Falls post office was closed after seventy-seven years, and in 1935, Trinity Lutheran Church of Deerfield sought to merge with St. John's Lutheran Church in Owatonna. Trinity had been served by the pastor from St. John's, and on January 11, 1936, the thirty-two families voted to close their church and join St. John's.

A Changing America and a Changing World

In April 1930, Dr. M. A. Hanson, a veteran of World War I and a major brain surgeon, spoke at an Owatonna service club and warned the audience that "we should expect a horrible war within a generation because of what communism was doing to the yellow races of Asia." But at the time most Americans were more concerned within the nation than in the world. In 1933, the National American Legion organized a campaign to reduce government expenses "by eliminating all those in government who provided no productive service." A month later, an article stated that thirty-three from the county had joined the Civilian Conservation Corps (CCC), twenty-four of whom came from Owatonna. They were paid $30 a month, $25 of which was sent to the parents. Walther Zatochill, a long-time employee of the Drache Truck Line, commented that when he and his brother joined the CCC their mother received $50 a month, "which was more than she had made working full time, but she continued to work because that was her nature." In 1934, the county was called on to fill twenty CCC vacancies for the state, and Eldon Halverson became the first from the township to enlist.

Farm numbers peaked at 6,454,000 in 1920 and then declined to 6,295,103 by 1930, but some key people in the USDA believed that small, owner-operated farms was still the way agriculture should be run and convinced political leaders that there needed to be a rural rehabilitation. The typical farm in southeastern Minnesota in 1930 was a 120-acre livestock operation. By 1935, under the farm rehabilitation program, farm numbers had rebounded to 6,815,103, an 8.3 percent increase in five years. Some of the reasoning for keeping people on the farm was that welfare was less costly for farm families than for urban dwellers, but it magnified the farmer's plight. About 1938, the industrial sector started to recover, and by 1940 farm numbers fell to 6,102,417, which was a drop of 10.5 percent in five years, and the trend continued.

In 1936, the first Social Security payments were paid. They started at $30 a month for those sixty-five and older and "must have been a citizen of the United States for twenty-five years and of the state for two years. Could not own property in excess of $3,500 or disposed of it within the previous two years." It was estimated that 26 million would be eligible to collect benefits by January 1, 1937. By November that year the county had received $1,215,762.99 for Old Age Assistance, Aid to Dependent Children, and Aid to the Blind. The 228 in the county who received Old Age Assistance received an average of $15.94 per month vs. the average of $19.54 for the state.

In 1936, the *Journal-Chronicle* joined the United Press network, and world news became front page news. The Spanish Civil War took the place of what was happening in Owatonna. Other articles on the first page stated that the Japanese were going to expand their submarine fleet and the atrocities of the Ethiopian War. President Roosevelt forecast that the deficit would be $2 billion. In March 1937, the Minnesota legislature was considering a state income tax that would be a maximum of 25 percent for individuals, which would be 66.75 percent higher than any other state. By October, Japan let it be known that it would not let fifty-three other countries interfere with her efforts in China.

An editorial in the *Photo News* eluded that big business seemed to be the major reason why the recovery was not moving faster, but it stated that government competition, strikes, and the growing federal debt were the problem. It closed "Government 'Help' to Business [is] Anything but an Aid Toward Recovery."

In 1939, communist party members in government grabbed the headlines. Dr. Bryn Jones, of the Kellogg Foundation, speaking at a service club luncheon, was quoted: "Every American has a Grave Stake in [the] New World War. Russia was not fighting, as a Big Winner. [He warned] that the winner of the German-Russian [conflict] will be a threat to America. The threat of communism is now greater than ever before in history."

One of the first responses to all the scare talk of war appeared in May 1940 when Owatonna Tool Company (OTC) marked its 50th anniversary. OTC was heralded as being one of the area's largest industries serving both national and international customers. It provided a direct living for 250 in Owatonna plus 125 salesmen. Its major customers were John Deere, Minneapolis Moline, Oliver, Cleveland Tractor, and Allis Chalmers, who were served from OTCs branch warehouses in four states in addition to Minnesota. It had customers in twelve countries on four continents.

In September 1940, the second response to world activity impacted the county when 2,471 men between the ages of twenty-one and thirty-eight were required to register for the draft. Arnold Abbe, son of the Alvin Abbe, was driving truck for Drache Truck Line and was one of the first to receive notice that he would be drafted for one year. I remember his words when he got the notice. He told me, "I wouldn't mind going in for a year but I have a feeling that it will be longer than that." Arnold was correct, for he was in almost five years.

In November 1940, eight nurses from the Owatonna Hospital who were members of the Naval Reserve Nurse Corps were ordered to active duty. Then, the National Guard units received notice to move to Camp Claiborne, Louisiana, along with units from Albert Lea, Austin, Jackson, and Northfield. Guard members wore their uniforms to high school, which gave fellow students notice of what was ahead for many of us. The entire citizenry received notice of what might be ahead when the food stamp system was put into effect.

Then on Sunday, December 7, "Japan served notice upon the Americas (by attacking Pearl Harbor) and the world that all the Axis forces must be destroyed if any portion of the human race is to have peace and if any people is (sic) to have security from ruthlessness, destruction, and hate." Colonel Harold Nelson, director of Selective Service, reported that dependents would no longer hold men from the service. He said, "We are rapidly approaching the stage where everyone must be working or fighting to win the war." Six months after graduation from OHS in 1942, 71 percent of all males who were physically fit were already in the service or had volunteered and were waiting to be called up.

The President requested that the speed limit be set at forty miles per hour to save rubber and reduce the danger of driving with well-worn tires. The limit was later reduced to thirty-five, and gas rationing was instituted. To save metal, license tabs about 2x4 inches that clamped over the plates were issued. This also reduced the large, heavy paper envelopes for mailing them. Marigold Dairies was ordered to reduce its production of ice cream to 65 percent of their 1942 output. They could not take special orders because they had to make sure they could supply their regular dealers. Jostens built a large addition to handle defense contracts, which required 85 percent of their master mechanics' time. By then 60 percent of all their employees were working on secret precision military weapons.

War Loan Drives were a part of the government's effort to finance the war as it was being fought. For the first time women had the responsibility for chairing the campaign. Meriden Township, under the direction of Gladys (Mrs. Arthur) Uecker, had a quota that was $6,000 higher than any other rural township, but that only challenged them to do more. They raised $63,046.66, which was 105 percent over quota. In the January 1944 Fifth V for Victory Bond Drive, the county quota was $2,100,000 and the township goal was $107,520. Gladys Uecker chaired the drive and was the first woman chair to go over the top in both bond and individual quota with a total of $136,539. Uecker headed the sixth and final drive, and again the township topped all others in the county with $105,702.98. Probably the only Meriden civilian who had a more active part in the home front was Harry Andrews, who served on the Selective Service Board from 1930 to 1946. His wife, Lucille, told him, "Get off the board and quit having to listen to all these fathers trying to keep their sons out of the military service." World War II had far more impact on the lives of people in the township because a far greater portion of the population spent time in the service. Instead of just going to France, they went all over the world. Farming changed more than it did in the previous wartime period because of greater changes in technology. Probably the greatest change was caused by the veteran's benefit program from short courses for the older veterans and the great increase in numbers who earned college degrees.

Anna L. Drache, A hand-written memoir of her life in Meriden, written in 1970; Charles Johnson, letter, 7 July 1992; Proceedings of 4 April 1939 hearing, held at the Meriden depot by the attorney for the Railroad and Warehouse Commission of Minnesota; D. M. McCoughlin, Postmaster, letter to Honorable M. I. Ryan, Inspector in Charge, St. Paul, 10 November 1942; Drache, *Challenge* 270, 271, 276; Meriden RTC; Mildred Fette interview; Palas interview; Gladys Murray, several conversations; Born interview; Charles Willert, personal interview, 2 July 2009, hereafter Charles Willert interview; Paul Drache, Day Books, January 1, 1937 to October 6, 1989; Henry Steele Commager, "The Agricultural Adjustment Act," *Documents of American History* (New York, 1958) 422; Drache, *Furrow* 247, 335; Baker, O. E. "Farm Youth, Lacking Opportunities, Face Difficult Adjustment," *Yearbook of Agriculture*, ed. Milton Eisenhower (Washington, 1934) 207-209; Wheeler, McMillen, *Too Many Farmers* (New York: William Morrow & Company) 37–41; Max Drache, Farm Record Book; *Chronological Landmarks* 53–54; Clarence and Alice Eggers, personal interview, 9 September 2002, hereafter Eggers interview; Judy Ellingson Eggers, telephone conversation, 20 December 2011; Bernice Scholljegerdes Jensen, telephone conversation, 20 December 2011; James M. Andrews, letter, 25 May 2005, hereafter Andrews; Drache, *Plowshares*, 65, 97; Arthur H. Uecker, audio cassette, 20 July 1999, hereafter Uecker; Harold Severson, *The Night They Turned the Lights On* (privately published, 1962) 137–139, hereafter Severson; Edward L. Schapsmeier and Frederick Schapsmeier, *Encyclopedia of American Agricultural History* (Westport: Greenwood Press, 1975), 302.

CHAPTER VII
The Peak Years
1945–1959

The Impact of World War II

After World War I there was a popular song entitled, "How Ya Gonna Keep 'Em Down on the Farm (After They Seen Paree)," an obvious reference to the doughboys who had grown up on the farm and in the service were exposed to the urban world. World War II had a greater impact because a far larger percentage of our population was involved for a longer period of time and in the entire world rather than just Western Europe. In January 1945 the government established a ceiling plan to control the number of employees local firms could have available to shuttle workers to essential war work, which indicates that no one knew how much longer the war would last. Then Germany surrendered on May 7, 1945, and Japan on August 14, 1945.

In October 1944, to comply with the demand for more powdered milk, the Creamery Cooperative Milk Association in Owatonna had received its first whole milk in ten-gallon cans trucked in from Clinton Falls, Crown, Deerfield, and Meriden creameries. Within four months another eight creameries were sending their milk to the plant. Owatonna Milk Processing had produced nearly 5 million pounds of milk solids for the government in the previous year and was preparing to increase their production.

The government wanted more whole milk and paid a greater subsidy for butterfat in whole milk than it did for butterfat from cream because they wanted more skimmed milk for powdered milk for the military and the lend-lease program. This put the local creameries at a disadvantage and caused them to lose their farmer patrons, but it greatly increased the volume of milk that had to be delivered each day. The Meriden Creamery patrons had to buy the extra ten-gallon cans and put two trucks into service. They also contracted with a trucker to haul milk.

The Borden plant, in Owatonna, accepted cream from Steele County creameries and contracted with Meriden creamery, which still accepted cream to make into butter. The building was enlarged to handle both cream and the increased volume of whole milk. It was a profitable move as the expansion was paid for in two years. Meriden was making most of the butter for Steele County and shipped the skimmed milk to Owatonna. The creamery had six to eight employees in its final years. Much the same was happening in the other dairy counties in the area. Drache Truck Lines, which had started hauling buttermilk and skim milk in 1938, by 1945 was hauling from seventy-four creameries within a 100 mile range of Owatonna. Most of its fifteen employees lived either in New Ulm or Owatonna. The demand continued after the war because so many products not known to the retail markets prior to the war used milk by-products, including many processed foods.

The Drache truck fleet and drivers; three more semi-trailer freight rigs and a milk truck were added prior to 1946. L-R: Paul Drache, Wally?, Elmer Ackman, Jim Lennon, Herman Roeker, Floyd Ness, Zatochill.

In March 1945, the government drafted thirty-eight from the county. By October 1945, Steele County had provided 2,066 (this did not include women) to the service, of whom 1,460 were in the army, which included the Air Corps, 520 in the navy, 73 in the Marines, and 13 in the Coast Guard. As of June 13, 1946, 1,542, men and women had been released. On February 23, 1946, Arnold Abbe's bride, Janet Moffat Abbe, arrived in New York and became one of the township's war brides.

The war shrunk the world leading to what we now call globalization. The United States, now the world's leading power, had a major part in the United Nation's Food and Agricultural Organization, which was founded in October 1945. World hunger became an issue in which our agriculture has played a leading role. The military physical examinations gave the government a solid data base about the relative health of the American society. In 1946, one of the first moves to improve the nation's health and nutrition was to establish the National Food Lunch program.

The GI Bill, which exploded the number of people who received college degrees, led to an unprecedented growth in technology and a standard of living that was unmatched by any nation, as well as making us the dominant world power. Training courses for those desiring to farm had a major bearing on agriculture, which led to the Second Agricultural Revolution and made America the world leader in food production while much of the world was plagued with constant shortages.

The first mention of airplanes for people in the township was in 1919 when a plane circled the village and also when barnstormers toured the nation and had shows in Owatonna. However, after WWII airplanes became commonplace, and by 1948 more than 2,000 private planes used the airport. Many of them were spray planes for weed control.

Another off-shoot of the war was the all steel Quonset building, which had been adopted by the military and used throughout the world. In 1948, the Alexander Lumber Company started selling them in Owatonna.

The Business District

Telephones

Communication systems were some of the greatest recipients of wartime technological improvements. In February 1945, Northwestern Bell Telephone announced that it planned to extend and improve the rural telephone systems as soon as war restrictions permitted. There was considerable complaining at the annual MRTC meeting because of the poor service, but the officers were quick to respond, "these are war times and we are getting about as good service as any company in these trying times." Business in the village was so good that an additional line was needed, but no material was available. The final business of the day was to get as many members as possible to meet "after corn planting" and make repairs on the line. At the 1946 meeting, W. P. Jones, who had long managed the daily operation because the telephone was so critical to the bank, reported that the new line to Meriden had been completed, and all the cost of that project had been paid.

As soon as materials were available, Oliver Stendel, Kenneth Brase, Paul Klamm, and Harry Grandprey worked under the supervision of Ernie Fette to build or rebuild thirty miles of line. The crew found "some of the original line spliced with baling wire. In one case barbed wire with the barbs still on it was not correctly spliced but merely hooked to cause a connection." The work was all done by hand augers and ladders or shoe spikes to climb the poles. Many new customers were "more than eager to sign up for a telephone." But now, instead of twelve to twenty customers per line, the most was six, and there were no complaints that the rate was $1.50 a month for a rural line. In Meriden a two-party business line was $3.75, and a private line was $4.50. For the first time since MTRC was established in 1913, a patron could reach the operator instantly.

At the annual meeting in 1949 a motion was made to warn people "not to use the phone too long or too often. People will be given three warnings and then [their] phone would be removed. Phone calls were to be limited to three minutes."

In September 1950, the eight farm residences on the 469 line of the Crane Creek Telephone Company "were enjoying direct dial service." In February 1951, all six lines out of Owatonna to Meriden were converted to direct dial, and it became the first rural company with its 400 members to be entirely converted to direct dial. It was expected that another 1,200 members would be added within a year.

Telephone technology continued to improve, and in 1955 people in Owatonna were introduced to automatic answering equipment that which recorded messages and gave out information "without any human supervision." That year Owatonna installed its 5,000th phone, making it one of the top per capita telephone cities in Minnesota. The first phone was installed in 1884; it took sixty-two years to reach 2,500 and only nine to add another 2,500. In late 1957, Owatonna was connected with the Twin Cities and eight other communities with new long distance direct dialing.

The Creamery

Switching to whole milk caused a logistic problem because of the much greater volume that needed to be transported each day. By 1947, the creamery had four trucks with insulated bodies to pick up the milk from the farms daily. Several patrons quit dairying at the time because they did not want to purchase the can coolers or a bulk holding tank or construct better milk houses required by new state standards to qualify for grade A milk. The creamery charged to pick up the milk, but to cover its cost it set a minimum that "pinched the smaller operator."

Uecker, the Meriden butter maker, stated that there was a significant improvement in the milk quality and in the end product. The creamery gained new patrons because Crown, Deerfield, and Golden Rule creameries had closed. In 1949, it switched to using natural gas for steam generation, which greatly improved efficiency. This also provided more room in the creamery because it no longer had to store the coal. In 1955, a truck was purchased with a 1,800-gallon bulk tank just to haul skim milk to the Borden plant in Owatonna. In 1956, the Lemond creamery closed and its patrons joined Meriden.

In March 1957, Uecker and others investigated the idea of six local creameries forming a larger co-op and expressed the feasibility of merging with Bordens. Meriden continued to grow and in 1958 hit a new high of 21 million pounds of milk and 807,958 pounds of butter; it sold its skim milk to Bordens. In 1959, the board debated the wisdom of accepting any more patrons because the facility was operating at its maximum capacity. However, the farmers were no longer coming to Meriden to deliver their cream or milk, which greatly reduced business in the village.

Trucking

In May 1946, upon the advice of his accountant and his banker, Paul Drache incorporated the trucking company. The business was changing sharply. By 1946, livestock shipments had declined to twenty-two loads, but the freight business required six trucks in addition to the bulk milk trucks. I had always thought that I would join Dad in the business after graduating from college and in June 1947 went right to work for the company, but in the fall of 1947, I decided to return to college to obtain courses required to teach. I remember the time and spot where I approached Dad and told him my plans. His instant reply was, "Then I will settle down to my best men and the best customers and get rid of the milk business." He had anticipated that the glamour of trucking was not to my liking and did not appear to be the least surprised. I remember how decisive he was and yet was so relaxed. He had buyers who were in the milk hauling business and wanted to expand. He sold at once to the Strohschein Brothers, who were former butter makers and very familiar with the dairy industry. Dad was surprised because the square, locally made bulk milk tanks that were acceptable under war time conditions no longer met standards for the industry and had little alternative value. Eventually, they were all sold for septic tanks, but I never knew that until years later when he was in the nursing home. The Strohscheins purchased the business, "a little blue sky," and three of the eight trucks. Dad realized early that if the freight deliveries were to places where backhauls were not readily available, he had to get backhauls. He liked the new challenge.

A Chronology of Events in the Business District

In August 1945, Henry Wicklow, who had been in the Meriden grocery business since 1895, decided it was time to retire. Wicklow ran one of the more successful businesses in the village and was definitely one of the stalwarts of the business district. The Wicklows were always involved, but rarely in the forefront of community events, including being on the bank board for several decades.

Going out of business was not entirely a pleasant event for him, for he had a series of going-out-of-business sales over a period of time in the hope that he could reduce his inventory. Anyone who traded at his store in the final decade of his business realized that much of the inventory on the shelves of that sizable building was very dated. Each week he lowered the prices in an effort to move the remaining inventory, and each week he personally hauled another truckload to the Owatonna city dump. He confided to Paul Drache disappointed he was him about how much he was throwing away because "people were just not interested." He did not realize how dated the shoes and clothing material were. Dad always had an eye for "old stuff" and purchased a pair of narrow-width ladies high button shoes and gave them to my wife; she wore them at events that depicted styles of the past.

After the unsalable inventory was disposed of, the store was sold to Mr. and Mrs. Leslie McCray, who owned the business until 1946 when Diemer and Lillian Davis purchased it. Davis was a proactive business man and remodeled the store and added fresh and frozen fruit and vegetables with a limited assortment of dry goods, shoe laces, and drugs. But he worked hard for the poultry business and, unlike most Meriden merchants, he did considerable advertising in newspapers, on the radio, and by direct mail. The Davis store was a busy place, and Diemer and Lillian, their son, Wayne, and at least another full-time employee were present six days a week.

With the increased activity in the business district some of the women decided that the village needed a place "to get a bite of food." In 1947, Anna Drache, with encouragement from Rosina Schultz and Helen Schoonover, financed the equipment that went into a lean-to on the Drache garage. It was named M & M for Meriden, Minnesota, and was referred to as the coffee shop where hamburgers and short-order meals were served. Rosina Schultz, who was the first to manage it, said there were periods when many meals were served at noon to various crews for the pipe line or for the Birds Eye Cannery in Waseca. "There were customers at the blacksmith, elevator, or lumberyard, or just passing through, so there was someone there on a steady basis." Rosina added, "I really enjoyed those years. The only time it was spoiled was if some one tried to bring in liquor." Rosina Schultz, Helen Schoonover,and Mildred , with help from teenagers took turns, operating the coffee shop until November 1951 when the food inspector informed them that all stainless steel sinks and other appliances were required. It was decided that the cost would be prohibitive and to closed the doors.

In 1947, when I decided to leave the truck line, I gave up my stock in the corporation and took over the Drache Garage. I rented it to Corwin Kanne for a brief period, and in 1949 to Herman Zacharias, who operated the business until 1955 when he purchased it and renamed it Zacks Garage. In the early 1940s, Ernie Fette erected a building on the lot between the big Fette house and the Drache house, which held the post office, to provide a place for his electrical business. He also opened a tavern that he and his wife, Evelyn, ran until Ernie was employed in Owatonna and they moved there. Then, the business was taken over and expanded by Murrel and Isabel Kleist. They operated it until 1952 when Walter and Gertrude Voss, who had discontinued farming, purchased the business and also the Fette house. They converted the house to two apartments and added a barber shop to the east side of the tavern. Harvey Moore came out from Owatonna one day a week to barber. In addition to the tavern, Gertrude Voss had a full-scale restaurant, two pool tables, and a juke box while Walter drove for Drache Truck Line. Ray Stuart was one of the regular customers for meals and gave the place the catchy title of Voss' Vitamin Vineyard, which was often referred to as the 3Vs. In 1959, after Harvey Moore quit, Gay Johnson barbered there for several years.

Anna Drache in the post office in 1953; when she became postmaster in 1929 a wanscot wall separated the parlor and the office, which measured 6 x 10 feet and was part of the home until 1969. The complete office is now at the Village of Yesteryear.

The Davis Store became a member of the Associated Grocers of Blooming Prairie, an association of 150 village grocers who bought as a group in an attempt to remain competitive with the chain stores in the larger communities. Davis did so much mail advertising that on July 1, 1953, the Meriden post office was raised to third class. When that happened the postal department mandated that office hours be established at 8:00 a.m. to noon and 1:00 to 5:00 p.m., and Saturday from 8:00 a.m. to 1:00 p.m. I can attest to the fact that my mother really appreciated the stricter regulation, but there was still considerable business done out of the family entrance because some of the old timers could not break the habit of getting the mail at their convenience. The cut-off point for a fourth class office was $1,500 in stamp sales, and the Davis business put it well over that. The volume for third class offices ranged from $1,500 to $8,000, but the Meriden business never exceeded $3,000.

In 1954, Leo and Dorothy Thompson purchased the store on the west side of the county road, which had been operated by six different owners since 1936. The Thompsons had a lively business for the next decade, in part because they served hamburgers and snacks as a sideline. In 1958, in addition to the business in the village, they opened the Le-Do Drive during the summer months. This business was located in a schoolhouse along Highway 14 at the junction of the first township road west of the Meriden junction, in section twelve of Woodville Township. That business was open for a few summers. The Thompsons also added the Doughboy Feed line and held feed and pork demonstrations that featured pancake and sausage feeds that drew as many as 300 people to the village.

Henry Olson, who had established a successful elevator business in 1937 as the Meriden Grain Company, retired in 1955 and sold the business to the Owatonna Farmers Elevator Company, which was managed by Duane Miller. The Millers were a positive influence in the village during these years when Meriden was thriving.

In 1948, all was going well in the First State Bank of Meriden, and W. P. Jones purchased the bank in Medford. His son, Ferris, operated it with the aid of Lyle Jones, no relation. Suddenly, in 1955, Ferris Jones left Medford and was named cashier of the Farmers National Bank in Waseca, and W. P. Jones assumed the leadership of the Medford bank. Alfred Schuldt, the long-time employee at Meriden, kept the day-to-day operation going in Meriden. In 1958, First State introduced "Bank-O-Medic" monthly deposits for those who signed up through the bank to carry hospitalization coverage with North Central Life.

In the late 1950s, the bank experienced a loss in assets each year, and in 1959 it suffered an operating loss. Jones was prepared, for he had observed that many small-town banks were having trouble and decided to move the bank to Owatonna. The decline took place as follows: from

1954 to 1959 time deposits dropped from $637,656 to $343,779.87 and mortgage loans from $679,850.95 to $323,951. It those years 341 savings accounts totaling $383,321.92 dropped to 201 accounts with $149,370.12, and 168 checking accounts with $185,209.71 dropped to 131 accounts with $162,500.04 on deposit. This was caused in part by the fact that the creamery had shifted its account to a bank in Owatonna. Many of the patrons shifted their checking and savings accounts at the same time. Stockholder dividends were $5,029.08 in 1954 and in 1958 had dropped to $679.46. In 1959 there was a deficit of $402.44.

At the Minnesota Commerce hearing, nearly forty individuals from the village appeared to object to the loss of the bank. Jones was the only witness for the bank. He testified that when assets started to decline, the directors decided to advertise on the radio. When other banks paid 1 to 2 percent on savings, Jones increased the rate to 2 percent and 2.5 percent on time deposits, and money came from many towns in the area. Then, Savings and Loans began bidding on saving accounts, but Jones found no reason to increase deposits because there were not sufficient applicants for loans. He continued his testimony by saying that since he had come to Meriden the stockyard was dismantled, the depot was closed, and the elevator had become a branch of the Owatonna Farmers Elevator. At first, farmers delivered their cream and deposited their checks in the bank and shopped in the village. When they switched to selling milk, the truck drivers delivered the checks to the farms. Improved roads enabled farmers to drive to Owatonna and Waseca to do their banking and shopping.

When the commissioner asked Jones what he intended to do, he replied that he was going where the depositors were or they were going to quit. The commissioner responded that he felt that there would be little opposition from the Owatonna banks. Art Willert was the only person at the meeting to speak, and he protested the move. The application to move was signed by W. P., Ferris, and Grace Jones, Lloyd Grandprey; and Paul Drache and approved by the Department of Commerce on December 1, 1959. In a *Daily People's Press* news article that followed Jones was quoted, "The decision to move was in the pattern of other farm community business, which migrated to larger cities through the advent of better roads and faster transportation." This event was an omen of what was to follow within a short period of time.

The Second Agricultural Revolution

The Technological Evolution

If a person who had lived in Bible times could have returned in 1830 they would have been able to farm because the equipment was basically the same. But if that person had returned in the 1890s, they would not have understood how to farm, for in the 1800s a revolution commenced that changed the world of farming. In 1814, the first iron plow with replaceable parts was developed, which was the key to better tilling of the soil. That led to the development of corn and grain planting equipment. In 1831, the development of the reaper led to better harvesting equipment. Every invention led developers to look for ways to further perfect the previous invention. All of the machines needed to be operated faster, which meant the plodding oxen had to be replaced by horses. Then, in 1892, the tractor was introduced, and after just over a century the horse age ended. By the 1920s more efficient tractors led to another revolution, which reached its full stride in the 1950s and has continued since.

The number of farms nationally peaked in 1920, and then the long decline set in except for a brief period in the 1930s when a federal rural rehabilitation program led to the revival of 500,000 farms. Most of those farms were abandoned as soon as economic conditions improved. When demand for food and fiber increased during the 1940s, farmers reacted by adopting technology at an ever increasing rate.

Thanks to the vision of the members of the Gavel Club, Owatonna was well diversified and kept growing. Its machinery manufacturers and food processors needed all the workers that the agricultural revolution freed. The city retailers were very alert to what was taking place. A news article in the October 14, 1946, *Photo News* read, "Unable to stand up under a weekly session of complaints from their rural customers the retail men have taken another vote." Instead of closing at 5:30 p.m. on Saturdays, they agreed to remain open until 9:00 p.m. Farmers might have stuck to the hours of the past, but as their industry became more mechanized and capital intense, they adapted to a changing life style. In 1949, the Farm Bureau put on an intensive campaign of forming farm families into hospital insurance groups under Blue Cross/Blue Shield. That campaign also included many people in the village and the township who were not farming but were retired or worked in Owatonna or Waseca.

In 1952, Paul Buecksler was named chair of the Steele County Soil Conservation District, and Larry Ruehling, assistant county agent and township farmer, cooperated to lead the Straight River Crane Creek Watershed project, another step toward its conclusion. As agriculture capitalized, farmers became more concerned about working every acre of their farm. That year another mile and one-half of drainage ditch was dug, and twenty miles of tile were laid, which had a direct impact on thirty farms east and north of the village.

In 1958, officials from Steele and Waseca counties met with area farmers to discuss channel improvement and the need to reduce erosion caused by tillage and the diversion around Goose Lake. It was then determined that the cost should be allocated 55 percent for flood control with the balance to conservation and wildlife. By 1959, the watershed covered 104.2 square miles and drained 63,713 acres of which 35,463 were in Steele,(mostly in Meriden Township) and 31,250 in Waseca. This showed how important the project was to Meriden farmers.

The 1950s was a low period in national agriculture, and some farmers looked for help. The Farmers Union was established in the county, and in December 1953 forty Deerfield and Meriden farmers formed the M & D Farmers Union Local. Art Croft, of Deerfield, was elected president; Weldon Beese, Dallas O'Neill, and Elmer Krause served as directors.

During the summer of 1954 the Minnesota Farmers Union held a meeting at the high school and the state president was the featured speaker. I went to hear what he had to say and to observe how other farmers felt about farming. When the speaker stated that their goal was to establish the price levels at 110 percent of parity, I realized how unrealistic that was, and it was a long time before I attended another farm organization meeting. It was during these years that banks learned that if they were going to finance modern agriculture, they had to have someone on their staff who was trained in that field.

Farm numbers continued to decline, but the importance of agriculture was greater than ever. Station KDHL 920, in Faribault, became the major voice of agriculture in the area, and in 1955, to strengthen its position in Steele and Waseca counties, it opened studios in Owatonna and Waseca.

A 1957 *Kiplinger Magazine* quoted a speech by its editor, "The *Farm Bloc* long a potent force in American politics will be an oddity of the past referred to only in history books 25 years from now. . . .There will be no such thing as farm states in American politics by 1982." Kiplinger continued that the *Farm Bloc* would share the view of the industrial states. An editorial in the *Photo News* in February 1958, taken from another source, stated, "Today there are proportionately fewer, by more than half, than there were in 1929 when 22 percent were employed in agriculture, now there are only 10 percent. If that 10 percent is to gain a favorable income the process of migration [off the farm] must continue. This makes farming. . .sound complicated and difficult. Well, in terms of making a good living, what isn't complicated and difficult?" The editorial closed on the note about the expertise required to be a good farmer.

Upgrading Agricultural Production

In 1946, Roy Bakehouse was president of the Southern Minnesota Breeders Federation, which had serviced 7,855 cows, an indication how quickly farmers adapted. (The Federation only produced semen for breeding and is not to be confused with the testing associations). In 1947, Leonard Gabbert, who had farmed near Moorhead, purchased the Gilkey Farms for $170,000. Gilkey had clearly been the most prominent registered breeder in the township and attracted considerable attention but not many followers among local farmers. Gabbert merged forty-five of his registered animals with some of the Gilkey herd, including the top producer, with a record of 29,616 pounds of milk and 1,200 pounds of butter fat. But things changed rapidly, for in 1950 some of the top registered herds in the state, including Gabbert, were losing their market; those who were not purebred producers continued to adopt artificial insemination. By 1952, 59,526 cows were serviced artificially as farmers realized that the least costly way to increase their income was to have better producing cows, not more of them. The Federation faced a challenge because the number of purebred herds that provided bulls was declining. At the same time the number of dairy farmers also dropped sharply.

What happened in the cow testing association explains why the number of dairy farmers was declining. In 1947, the twenty-six herds in the Steele County Cow Testing Association averaged 10,759 pounds of milk and 368 pounds of butter fat. The demand for milk continued to grow because of all the by-products that were being produced, but farmers rose to the challenge. By 1955, many herds averaged over 400 pounds of butter fat per cow with many top cows doing in excess of 500 pounds, which indicated that there was still room at the top. One cow of that group produced 18,423 pounds of milk and 681 pounds of butter fat.

The vocational agricultural courses at the high schools were having an impact. Not every farmer's son "was willing to farm just like Pa used to." This was a phrase I heard many times when interviewing bankers who worked with farmers not willing to change their method of operation.

In 1955, Don Wilker made news when he was a candidate for FFA Star Farmer award. By the time he was a junior in high school he had a net worth "of nearly $8,000 based on a farm program containing twelve dairy cows, twenty-eight hogs, and five head of beef cattle." Wilker had records to prove how he had accomplished that. In 1956, he was one of the first in the township to become an FFA Star Farmer.

In 1952, county agricultural agent Russ Gute addressed the annual Meriden creamery meeting. He described the direct relationship of cow testing and butter fat when having cows grazing on grass to using nitrogen when raising corn, insulating the milk house, and drying hay and improving feed grain before buying bulk milk tanks. The second speaker at that meeting spoke on the need for the American Dairy Association to do more advertising to inform people who were becoming calorie conscious at the same time that medical people were recognizing value of fluid milk.

While many farmers worked to improve their dairy facilities and increase their herd average, the Andrews Brothers realized that to support three families they had to have a much larger dairy than they cared to. However, they saw the national trend to large-scale poultry farming and opted to remodel their dairy barn into three floors for chickens, which they raised under contract with Swift & Co. C. C. Jolly, of Owatonna, enlarged his business by expanding into manufacturing poultry equipment. He moved Jiffy-Way manufacturing, which he owned, from Florida to Owatonna to gain greater efficiencies and be nearer the poultry business. Russ Gute advised that if farmers were going to depart from having "range free" flocks to raising them in confinement they would have to vaccinate their hens. Agriculture continued to become more management intense as farmers sought to improve their income.

All was not in vain, for Gute, who again spoke at the annual creamery meeting in 1959, discussed how the rising efficiency in dairying, pork, and poultry was greatly improving labor productivity. This meant that farms could grow with less labor, but Steele County still had 500 farms under 100 acres.

The second speaker that day was Al Camp, the long-time operator of Gilt Edge creamery in Owatonna, who spoke on quality control. He pointed out that too much of the county milk production was still under grade to make grade "A" standards. New research data indicated that it was beneficial to have the cows milked in a different location from where they ate and rested. The DHIA expanded to testing 695 cows. Odell Knutson had the top herd in the association.

Leonard Gabbert sold his registered dairy herd and concentrated on raising purebred Hampshire hogs and in 1954 received $4,150, "the highest price ever," for a boar pig of that age for any breed. The buyer was a farmer from Indiana. Levern Wilker was one of twenty-four farmers honored by the University's Swine Honor Roll sponsored by *The Farmer.* He averaged nine and two-tenths pigs marketed per litter, and farrowed pigs three to four times per year. The state average was seven pigs sold per litter with eight sows, with fifty to 100 pigs sold annually, while Wilker marketed 400 per year at an average weight of 214 pounds at 174 days of age.

In 1946, the Steele County Crop Improvement had fifty-six test plots of which three were in the township on the Levern Wilker, Roy Ebeling, and Ted Knutson farms. Gute announced that the farm test plots indicated that phosphate increased the yields of small grains and corn.

Cliff Richardson started a hybrid seed corn business in 1941 and handled about one carload of corn that year, but by 1949 his business had a volume of twenty-five carloads. The company had 225 farmer dealers in Minnesota and part of Iowa and Wisconsin, but Richardson had the only Pfister Hybrids office in Minnesota. They also handled a full line of hybrid seeds, fertilizers, insecticide, and sprayers. The firm required twenty-six seed corn growers to raise 1,000 acres of seed to supply their demand.

In 1949, the county was invaded by the corn borer, which affected about 50 to 90 percent of the crop. Aerial sprayers were rushed into service to apply 2-4D because it was the only chance of saving the crop. The Southern Minnesota Aviation Company, of Owatonna, had a good workout trying to keep up with the demand when soon after there was a grasshopper infestation.

Aerial spraying had become popular almost immediately after the war in the cotton growing areas, but 1949 was the first year of widespread use in the county. In 1952, planes were used for soil mapping, and farmers were provided with flights so they could view the damage caused by poor conservation practices. That year the first county plowing contest was sponsored by the Soil Conservation Service in an effort to show the need for better tillage practices.

Minnesota had a record 232 corn farmers who established new records in corn production. Included in that group was Orlo Sette, of Owatonna Township, who placed second in the Minnesota Xtra yield contest with 150.2 bushels on a two-acre plot. The state winner from Kiester had 160.9 bushels. By 1951, the improved varieties had caused an increase to 68,000 acres of corn in the county, but dairying was still important, so 58,000 acres of oats were planted. Oats yields reached 102 bushels plus forty-six bales of straw, which was almost as important as the grain for dairy needs. Soybeans lagged at 11,300 acres, but as dairying declined, soybeans quickly filled the void. An article in the *Photo News* in March 1957 was headlined "Plant Corn with Little Tilling in the Tractor Track." This was an early attempt by Gute to explain the advantage of minimum tillage. He stated, "This is a new system that's gaining popularity. Disking fields for corn may someday be obsolete in many areas of Minnesota. It also avoids packing the soil."

Mechanization Replaces Manpower

There are two stories about tractors that date to the 1930s that are firmly implemented in my mind because they involved two people I knew well. I inserted them here because they show why agriculture continued to produce a surplus. The first took place in 1934 when Delbert Brase, son of Mr. and Mrs. Herman Brase, purchased an Allis Chalmers tractor with rubber tires. Rubber tires for tractors were first introduced in 1928, so many farmers were curious and some were pessimistic that they could replace the narrow-wheeled spade lugs in mud. Delbert was a bit of a showman and announced that he would demonstrate the tractor. Both the tractor and tire dealers were present when Delbert drove his tractor through a low, wet spot where a steel-wheeled tractor was hopelessly stuck. Delbert's brother was a classmate in grade school and we all had conversation for days about that rubber-tired tractor.

The second event took place on the Alvin Abbe farm in 1936 when sleeping sickness ravaged the countryside. I bicycled past their farm and noted they were pulling the binder with their 1926 Dodge sedan. There was a ten-gallon cream can full of water on each end of the field and a smaller can positioned on the front bumper. I knew that they had lost horses to the disease, but the weather was perfect for harvest and they were anxious to finish harvest before it stormed. The next day I traveled by that field and the binder was being pulled by a new I.H.C. Farmall.

I was very impressed with the rapid transition from horses to tractors, but it was not until later that I realized that sleeping sickness was a blessing in disguise. The tractor was a far more efficient source of power. The Abbes were good farmers, and they had the resources to purchase a tractor, but that was not the case with all farmers in 1936.

In 1938, Secretary of Agriculture Henry A. Wallace cited that one of the seven causes of the farm problem was that since 1920 so many horses had been removed from farming that it had freed about 40 million acres of land to produce crops for the market and not for feeding horses. In total about 95 million acres of land were freed during the process of converting to tractors. Len Mosher got rid of his horses in 1922 and quite likely was one of the first farmers in Steele County to be totally mechanized. He was featured in the *Photo News* in 1950 pictured with a Farmall M with a four-row mounted cultivator. The farm had 518 acres of corn and a total crop base of 1,080 acres, one of the larger farmers in the county. At that time the county only had 118 farmers larger than 260 acres.

In October 1948, the *Photo News* ran a series in which it featured area farmers. In one article a farmer who had been farming since 1929 was featured. The couple had a 120-acre farm with an eight-room house that was built in 1940 and contained piped air, electricity, running water, and telephone. The farming operation had three horses, a tractor, four milk cows, sixty-three hogs, and a flock of 700 chickens. The major crop was thirty acres of corn, which provided silage for the cows and corn for the hogs. The remaining acres were in alfalfa, wheat, barley, and oats. When I visited with this person I learned that he was not involved with any farm organization, did not attend any farm-related meetings, and virtually had no interest in what was taking place in agriculture and said as much. They both were content with farming as it was and I never heard that they ever complained about farming.

During this period sweet corn raised for the canning company was harvested for the first time with a mechanical harvester. A two-row tractor-mounted harvester traveled down the field and stripped the ears from the stalk and passed the ears to a wagon that was pulled behind the tractor.

In 1860, Owatonna Manufacturing Company (OMC) started making butter churns, then added grain seeders where the seed was dropped into a furrow made by a shovel opener. Later they made grain drills that placed the seed in the ground and covered it. In 1928, they developed the portable elevator. In 1952, OMC purchased the rights of the Ommodt self-propelled grain swather. The swather added a major implement to the company line and required a major expansion to the plant. OMC properly sensed the revolution that was taking place in agriculture and soon added more implements to its line.

Top-notch farm managers quickly learned that a computer could become one of their greatest time savers and most profitable pieces of equipment. In 1957, Levern Wilker secured the services of Electronic Farm Records, Inc. whose members had created Agrivac System of Farm Accounting. Wilker stated that the new system had reduced his accounting system time to one-fifth of his former method. It used a punch card system that could process 26,000 cards an hour and could process his monthly records in a matter of minutes.

During the 1950s, the county agent explained that artificial crop drying would soon be standard practice on the farm. "Drying is no longer just an emergency measure. If one uses a picker sheller [predecessor to corn combining], drying makes it even more feasible than ever to store corn on the farm." Farm machinery dealers realized that having an annual event with movies and speakers was an effective way to arouse interest in the technology that was appearing at a faster rate than ever. That event replaced the former annual creamery and/or elevator meeting and the medicine man show of previous decades.

In the previous chapter it was related how the impact of electricity came to many farmers in Meriden townships earlier than in many other townships because power lines ran through the area connecting the larger communities. However, the influence of the REA on rural America cannot be overlooked. By 1949, the Steele-Waseca Rural Cooperative had 3,900 customers in its ten counties. When the REA was first projected it was estimated that the average customer would use about 100 kilowatts per month, but by 1948 the average was 600 KWH per month. "The farmer has been able to do his work far more economically and in less time. . . .Electricity on the farm is a new way of living [and working]." By 1954, the cooperative had 1,600 miles of line and 4,500 customers. It had an impact on the rural area second only to the automobile. That change and many other innovations proved a real blessing for the farmers of the nation and its consumers.

A New Direction for Agriculture–Continued

In March 1949, a Congressional committee reported that in 1929 the federal government had 350 agencies and in 1949 it had 1,800. The article concluded, "It has resulted in a form of organization that is hopeless by way of economy." The war could not be blamed for all the agencies, but they were easier to create during the emergency than they were to close down after the war ended.

After WWII and the Korean War ended and other nations were able to increase their agricultural production, U. S. farmers were again caught with surpluses. O. B. Jesness, of the University Extension Service, wrote, "The price support program aggravated the farmer's plight in the long run because prices were high enough to price American farm products out of the world market, and they encouraged the better managers to adopt technology at a record pace."

Agriculture changed from a labor intensive to a labor extensive and capital intense industry and continued to increase production. The government reacted with the Agricultural Act of 1956, and during the decade the number of farmers nationally decreased by 31.1 percent and the number of people living on farms dropped from 23 to 15 million. Millions of acres were placed in conservation reserve, and many farmers who were not doing well were able to leave farming gracefully. Virtually none of them returned to farming. The township has top quality land and I am not aware that any of it was placed in the reserve, but farm consolidation was taking place rapidly because of the guarantees provided by the government. The bank became the township's first victim of those events.

In September 1959, the Food Stamp Program was made into law. It had a far-reaching impact on agriculture because of the amount of food allocated to the needy. Farmers did not like the fact that the cost of the program was included in the budget for agriculture but farm state legislators realized that this was necessary to get urban legislators to vote for it.

Daily Life and Culture

A Hodge Podge of Events in Meriden

The Grandprey twins, Lloyd and Loie, who were born in 1930, were the fourth generation Grandpreys to live in the village. Their parents, Lloyd and Eva, were both very involved in the American Legion and Auxiliary in Owatonna. Lloyd was well known in Owatonna because he blew "Taps" at funerals for decades. After the twins were old enough, I had the opportunity to be "baby sitter" many evenings during the years of 1936-1938, for which I was paid $0.25 per evening. I liked the job because the Grandprey house had "modern plumbing facilities" and they subscribed to the *National Geographic Magazine*. Their home was a short block away and no one ever gave a thought about walking the unlit streets in those days.

After I started OHS in 1938, Marvin Wilker replaced me as the milk delivery person, and when he went elsewhere he was replaced by the Grandprey twins. Lloyd Jr. told me they delivered milk until Paul Drache stopped selling milk in 1948 when the Davis store received pasteurized milk from Marigold Dairy. This was at the time of a polio epidemic and it was reputed that pasteurized milk was safer. Lloyd Grandprey, Jr. added that the twins put half of their pay in savings stamps and later recalled that he cashed in the bonds to pay for part of his honeymoon and part toward an engine overhaul on his car. When Paul Drache started selling milk in the village in 1928, he sold the milk for eight cents a quart, and when he quit selling in 1948 he was still charging eight cents. When I asked him why he never changed, he grinned and said, "It was just an accommodation, we had the farm and we needed milk so the delivery did not have to make a profit."

Lloyd Jr. said that they had many opportunities to keep busy in the village because several families still kept chickens and they had jobs cleaning the coops and also mowing lawns. Many of these were older people, but others were busy with their regular work. Farm boys would not have bothered to mention cleaning chicken coops because that musty job was routine for them. I think we all hated it. The twins also helped at unloading cars of lumber and stacking it in the sheds under the supervision of Art Willert, who worked at Hayes Lucas for many years. They also helped Willert build brooder houses for chickens or pigs during the spring and fall or whenever a farmer wanted a hay rack, pig feeder, and even outhouses, including a "three holer." When they were older they rode their bikes out to the farms to help make hay, harvest grain, and haul manure for fifty cents to a dollar a day plus dinner and morning and afternoon lunch. In their free time they did what the village youth had always done—they loitered at all the business places.

Starting in the 1940s some of them had newspaper routes, and for some Ray Stuart had special errands so he did not have to leave his studio. Eddie Kath worked in a café in Waseca during the day shift and drove cab in the evening in Waseca. He liked it when the Stuarts were in town and wanted to return to Meriden. Eddie said, "They would usually have a case of beer and some hard liquor that had to be carried up the stairs to their living quarters, but I was always rewarded with a $5 tip. That was big money for those days."

I recall that on Sunday afternoons during the 1930s, Norbert Abbe, Irwin Woker, Billy Frodel, and I rode our bikes out of the village and then returned to the pool hall and bought one quart of factory-packed ice cream. The four of us devoured it and then biked off in a different direction, or we would find a cool spot to sit down and visit. If it was hot and we had enough money, we would buy another quart of ice cream of a different flavor.

I remember a few Sunday afternoons in the winter when Art Zacharias drove his Model A and the same four of us would tie a rope to it and ski in the snow-filled ditches as he drove. The car did not have enough power to pull all four of us from a dead start, so we had to adjust our ropes so we could start in sequence.

The girls had similar activities—the outdoor movies each summer or ice skating on the pond on the low spot west of the creamery. Some recalled watching for President Truman's train coming through town in 1944, "But Meriden did not rate a stop like Owatonna and Waseca did." Most of the girls worked in the stores or at candling eggs and/or house cleaning.

Other than church and school, probably more people were involved in 4-H than any other activity. In 1947, two clubs were organized, which helped the county have a peak of 445 members. One of those clubs was the Meriden Happy Helpers, which had about seventeen to twenty-three members.

St. Paul's Church hosted a Brotherhood gathering of twelve communities totaling 330 members in May 1947 to hear Governor Luther Youngdahl speak on "The New Frontier in Moral and Spiritual Resources." He said, "The rising tide of broken homes, divorce, and parental and juvenile delinquency demands that Christian manhood of our state and nation arise to combat these evils and explore our moral and spiritual resources if our nation is to continue its glorious history or decay and disintegrate." Not much has changed in that respect.

The 1948 graduating class of the village school, which consisted of Marion Schoonover, Barbara Rietforts, Jerome Schroeder, and Charles Willert, made history for being the first class where every member graduated from college. That year the rural school graduation event was discontinued. The reason given was that the cost involved was rising, many teachers were already out of the area, and the state board of examinations sent out their marks with an admonition that students should continue their education.

The final event for the rural schools that year was an appeal for consolidation. Art Willert was on the county committee for that cause. In 1951, all the rural school board members gathered in Owatonna to get a report on the state's activity regarding consolidation. Each year since 1942 the percentage of rural students continuing their education increased. The hope was that consolidation would encourage more to do so.

On April 14, 1950, Shirley Sandborn, whose parents farmed along Highway 14, was killed as she alighted from the school bus. The driver of the car was charged for failing to observe the school bus sign, leaving the scene of an accident, and driving without a license (it had been suspended), and was fined $207. This was not the first traffic fatality in the township, but it was the first involving a school bus.

A happier event took place in September 1950 when Don Bakehouse, son of a long-time farm couple in the township, was elected as an officer in the National FFA. His father, Roy, received an honorary American Farmer degree at the same meeting.

The village upgraded in 1950 when natural gas was piped in, and the streets were black topped. Some homes had their own well and septic systems, but there were many cases where wells were shared.

The Meriden Grain Co., Davis of Meriden, and the First State Bank scored a big hit in 1952 when they sponsored the annual calendar painted by R. James Stuart. "The demand was widespread."

This photo was on the 1951 First State Bank of Meriden calendar. The depot was downsized after it was closed in 1939. This was the original village, minus ten homes on east Mill Street and six on the north along the country, at its peak even though it had already lost a couple of firms by this time.
Rosina Wilker Kopisckhe photo.

On January 23, 1953, the westbound Chicago to Black Hills Express hit a truck at the crossing in the village and took the lives of Marlow Schroht, 18; Arnold Voll, 25; and Harold Lewison, 51. Voll lived in Lemond and the other two were from Meriden. Skid marks indicated that the driver had spotted the train and had attempted to stop. A petition signed by 400 was presented to authorities to secure automatic warning signals, which ended in the decision to post stop signs.

In 1954, the Rural Fire Association purchased a tanker truck for the use of the Deerfield Mutual policy holders. This truck was stationed in the Owatonna fire hall and available to Owatonna residents. Better roads and much improved trucks made this a feasible move and greatly enhanced fire protection in the rural areas.

In 1955, a federal school milk program was introduced in all rural schools in which some schools paid a portion of the cost. The milk was delivered in ten-gallon cans supplied by the dairies and offered to the students twice daily.

In 1956, the Mother's Club of Meriden, under the direction of Mrs. Duane Miller, staged a program for the March of Dimes. About 100 attended to view two movies about polio after which lunch was served.

In February 1957, ten juveniles were arrested at a beer party on a farm near the village. Two boys, both age eighteen, paid fines and costs of $8.50 each to a justice of peace in Ellendale. "Some drinking may have gone on earlier in the home of the elderly farmer whose sixteen-year-old daughter was present. Eight younger children were referred to the Steele Count juvenile authorities. None were reported to be drunk."

The other event for the year was that the state decided to systemize the rural school districts to four numbers, and District 48 became 2118. See the school listing in the Appendix for the original and revised numbers of each district in the township.

Paul Drache owned ten acres in the wooded area of section five, which had provided fire wood for several members of the family since 1916, but after the village received natural gas he sold that land to William Heuer, of Owatonna. In the mid-1950s, Herman Zacharias, Luverne Zacharias, and Dale Lebahn purchased three lots and erected a saw mill to saw logs and build their houses. The added touch to their homes was that they purchased cherry wood, oak, and walnut logs. Others liked the idea, and by 1970 there were nine homes tucked into the woods. The village of Meriden had a suburb, and those who were still working were commuting to Owatonna or Waseca.

In June 1958, the Meriden Meadowlarks 4-H club had a program about the history of Home Economics Day (July 2) at Clear Lake Park after which they toured the Arnold Abbe and Martin Dinse homes and had a picnic dinner and wiener roast. Mrs. Lloyd Ellingson, the correspondent for the *Photo News*, always had a good column of social news but rarely ever wrote any business news. At the February 1959 meeting of the Meadowlarks, Kenneth Dinse gave a presentation on how to control parasites in pigs and Bernice Scholljegerdes gave the speech that she gave when she won first place in the County 4-H Radio Speech contest. That was a very successful year for both the Meadowlarks and the Meriden Skippers, for Janet, Hans, and Herman Hohrman won a trip to perform at the state fair talent show, and Judy Steinberg, of the Skippers, won the grand championship for the best Ayrshire dairy animal. Joyce Abbe won a blue ribbon in clothing, and Kenneth Dinse took top prize in the dairy judging.

During 1958-1959, the Mason Bus service operated teh buses to transport 380 rural students to Owatonna. In the following year all rural schools were expected to start sending their seventh and eighth graders to OHS. Because the village school had a different rating, it was permitted to delay sending their two top grades until 1963.

The year closed on an upbeat note for R. James Stuart when the advertising firm Thomas Murphy Co., of Red Oak, Iowa, offered to purchase all of his paintings for $500 each. It appears that he made a sale prior to his entering the county nursing home.

Beyond the Township Borders

A December 1946 *Photo News* carried an ad that Kottke Jewelers had the new ball pens for only $25, plus tax. Just months before the assistant cashier at the bank had proudly displayed a Reynolds pen that wrote under water and could be used to make carbon copies; it was only $46. He no longer had to use an indelible pencil. By 1952, the Paper Mate Ball Point pen was available for $1.69 at print shops, and within a few years they were used as advertisements by many firms.

In 1947, the county purchased a $14,950 SNOGO that could cut through a six-foot deep drift and throw eighteen tons of snow 200 feet a minute on a still day. If the wind was favorable it could throw the snow a quarter mile. This rotary with its 170 horse power engine was mounted on a 7 ton Oshkosh truck. The rig was delivered just before the February 1947 storm dropped sixteen inches of snow and train, bus, and private travel were all shut down. Within a few days all the county and rural roads were opened. Then another storm came and brought everything to a standstill once again.

The SNOGO at work entering Meriden from the south, March 1951.

By 1948, traffic had increased so much that Owatonna installed parking meters around Central Park and on all major streets as well as in all off-street parking lots "to make down town more customer friendly." Only four years later Central Park was made square so more parking spaces could be added and traffic control would be easier. "The wording in the original grant was 'a square' so those with a historical bent were finally appeased."

The high school announced the introduction of a driver education program ,and the fair grounds would be used for training purposes.

In the summer of 1949, single-room air conditioners "that could be easily installed in a window" were on sale for $345. "Pay only $34.50 down and the rest in easy installments." An 8.8 cubic foot home freezer by Frigidare could be purchased for $329.75 on easy terms. The most popular records were *Mule Train, Sweet Georgia Brown, That Lucky Old Sun*, and the *Merry Go Round Waltz*. If you wanted to dance, go to the Monterey Ballroom where Henry Charles played modern music one day; the next dance Wally Pikall and His Dill Pikals offered old time; and two days later the Six Fat Dutchman were playing. Those who did not want to dance could drive a half-mile north of Owatonna to the drive-in theatre.

The government announced that Social Security, which had started in 1935 by deducting one-half percent of wages up to $3,000 and one-half percent for the employer,would double that amount in the 1940s. In 1950, the deduction was raised from two percent to three percent. In 1959, there was a proposal in Congress that in the 1960s Social Security taxes should be raised every three years.

By 1949, total premiums for Federated Mutual Insurance had reached $12.7 million. It announced that its health, casualty, and accident insurance business was increasing so rapidly that it would construct a three-story building with basement which would double its buildings size. That provided for up to 650 people.

The rural areas continued to decline, and in 1955 the Pratt post office was closed as an economy move; in 1957, the Steele Center creamery closed; and in 1958, the Alexander Lumber Yard at Medford, which was opened in 1898, was closed because the company felt that it could offer better service from its Owatonna yard. But Owatonna continued to grow.

The railroads were still having difficulty in 1956 forcing the CNWR to sell 287 steam locomotives for scrap and discontinue eight passenger trains. Some transcontinental service was discontinued because it had a net loss of $5.5 million. In 1959, the road appealed to drop two passenger trains between Mankato and Chicago, which was allowed to happen in 1960. By that date the CNWR had lost most of its less-than-carload freight to trucks, but there was a bright spot, for it had a fleet of 710 diesel engines that were far more powerful and efficient. The Chicago and Rock Island, however, was experiencing good times because it had converted to A-frate, which involved seventeen-foot cargo boxes that carried all types of freight, including refrigerated units, which were placed on flat cars. It also improved its "auto-mobile" carriers. Both the A-frate and auto-mobile carriers were made into special trains that traveled much faster than the traditional freight trains. The cargo boxes were handled by fork lifts and quickly moved to the customer. Both of the above removed many trucks from the highways.

The interstate highway system, which was constructed during the 1950s, was designed to handle a high volume of traffic with minimum urban interference. New road construction methods were designed to handle the massive projects that were capable of laying a 2,000 to 2,600-foot long by 24-foot width of cement highway per day. I-35 was planned to reach from the Twin Cities to the Gulf of Mexico. It was supposed to have been finished in 1957 but was delayed because of a shortage of steel for bridges. In August 1958, the first section of the system in Minnesota consisting of 8.3 miles north out of Owatonna was completed. This was supposed to be "the safest highway in Minnesota," but the second day after it was opened the first auto accident took place when a driver fell asleep and crashed into a guard rail near Medford. Husband and wife were both hospitalized.

In 1958, OCCO purchased the pickle plant at Dodge Center. During peak season that plant had 125 employees, the Kenyon plant had 225, and Owatonna had 500. This was in addition to all the employees needed to raise the crop and get it to the factories.

Steele County became the first county in Minnesota to provide home nursing care for the chronically ill. This service gave nursing and rehabilitation service to any ill or disabled person in his or her home on the basis of need. About the same time the Owatonna Nursing Home was opened in what was heralded as a new era for elderly care. This facility replaced what was formerly called the county home.

Owatonna experienced a steady growth after WWII, and by 1957 employment reached 5,098, of which 1,631 jobswere in manufacturing, 1,028 were in trade, 770 were in service, 531 were in finance, and 496 were in government and education. The city had a severe housing shortage in the 1940s, but by 1958 1,300 new homes had been built, and in 1959 a new record was set with over $5.1 million in new construction.

The village hit its business peak during these years, but the seeds of decline had been sown. In 1939, the CNWR closed its depot. In the early 1940s the two Fette brothers had sold their businesses and moved to better opportunities in Owatonna. The First State Bank of Meriden experienced its first loss year and made the decision to move to Owatonna. The State of Minnesota gave notice that all rural schools would be closed and consolidated into the Owatonna district. Farm numbers declined, but agricultural production blessed the nation with its steady upward trend.

Chronological Landmarks 68-76; Meriden RTC; Minutes of the Meriden Creamery Association; *The Daily People's Press and The Photo News*; Drache, *Plowshares* 130–132; Drache, *Furrow* 46, 49; *Land of Plenty* 58–59; Robert Scholljegerdes, letter, 16 July 2002; Lloyd Grandprey Jr., several telephone conversations, 2008–2011; Marion "Loopsy" Zelinski, letter, 1 October 1997; Dale Lebahn, telephone interview, 20 June 2010; Edward Kath, several telephone conversations, 2008–2011; District Reunion Committee, "History of Meriden School District 48/2118," 25 June 1989.

CHAPTER VIII
A Period of Adjustment
1960–1974

The Business District

In April 1960, Diemer Davis, Leo Thompson, Harold Schroeder, and a few other business people felt that the village ought to be incorporated. Nothing was mentioned in the paper about what their reason for incorporating was, and there was little conversation about the issue. According to Gladys Uecker, who was the census taker that year, the village had 175 people. Fifty-one of them signed a petition opposing the movement and presented it to the county commissioners who moved not to take any action on the matter. The only outcome of that movement was that a few street lights were erected.

The Bank

As soon as the decision was made to move to Owatonna, Jones had a temporary frame building constructed so the bank could commence operations while a permanent building was being constructed. On March 29, 1960, the new board was made up of W. P. Jones, president; Lloyd Grandprey, vice president; with Grace Jones and Paul Drache, directors. That day the Jones family also sold their interest in the State Bank of Medford.

The name First State Bank of Meriden was changed to Oakdale State Bank of Owatonna, and the capital stock was increased to $75,000 divided into 1,500 shares of $50 each. The highest liability that the corporation could assume would be thirty times its capital and surplus. On July 9, 1960, a Drache truck was loaded with money and the bank records for the move to Owatonna under the protection of the county sheriff and a deputy. They were met by the Owatonna police at the city limits and guarded until everything was safely locked in the new bank. The Meriden Bank was formally closed on Saturday, July 11. On July 13, 1960, Oakdale State Bank was open for business as Owatonna's third bank.

In January 1961, the Oakdale State Bank had its first board meeting in the newly completed building. At that time there were twelve remaining stockholders with 252 shares of stock who had roots in Meriden. The other twenty-three stockholders held 1,548 shares of the 1,800 shares total, representing the capital stock of $90,000. By then the banks assets had rebounded to $596,410.58. In 1962, Alfred Schuldt, the assistant cashier since 1928, retired, and in 1963 Lloyd Grandprey and W. P. Jones both became sixty-five and had to retire from the board. At the same time, Ferris Jones resigned.

At the 1965 meeting it was announced the name would be changed to Owatonna State Bank. During 1964 assets had increased by 33 percent. Paul Drache had reached age sixty-five and retired, ending a connection with the bank that had started in 1917 when he served as assistant cashier. His departure was the bank's final connection with anyone from Meriden. In 1971, W. P. Jones was interviewed by the *Photo News* about a change in the bank and made a prophetic statement about its move to Owatonna. He said, "It was either die on the vine or move forward."

Telephones

On November 9, 1961, Lloyd Grandprey, secretary for the MRTC, sent a letter to all stockholders of the company that on December 12, 1961, its annual meeting would be held in the town for the purpose of voting on a resolution of the board that Meriden Rural Telephone Company should sell its physical assets to Northwestern Bell Telephone Company for the sum of $8,500. The board was determined that the sale be made, and if a two-thirds majority was not received, the board would submit the resolution to the court for the purpose of dissolving the company.

In 1913 when the first three lines were built into the township from Owatonna and one from Waseca, the monthly cost was $2 per month. It was not until 1959 that the rate was increased to $3.50, but some lines had as many as eighteen subscribers. W. P. Jones became manager and lineman in 1945 and continued in that position to the very last. In his final week serving as a lineman in the fall of 1961, he no longer lived in the village but had to climb ten poles to find any source of service interruption. "It was the second Sunday of hunting season. I got a call. . .telling me that the Meriden lines were out of service."

The article continued that he took Glen Koval with him, and a few miles west of Owatonna they found a cable where twenty-two wires had been severed. Twelve empty twelve - and sixteen-gauge shells were found at the scene. The problem was located at 8:30 p.m. and by 11 p.m. service were restored. The vandals were never located but they probably were people who was angry at Jones for transferring the bank.

Jones spoke of other problems that he had experienced in his years as a lineman and requested that he be relieved of his duties. The assembly expressed its appreciation to Jones for his work in keeping the lines in "tip top" shaped. Lloyd Grandprey had paid a $34.11 bill for long distance calls by Ray Stuart and it was moved that he be reimbursed by the company ,which had a balance of $209.50 at the bank. After all other issues were discussed a motion was made to vote on the sale. "Paper ballots were handed to the eleven stockholders and all eleven voted to sell the company." The final motion was that the company would pay for lunch for all present. The company had never paid a dividend, but after the sale the fifty remaining stockholders received a pro rata share of the proceeds.

The purchase agreement stated that all fees were paid to January 19, 1962, and service would be uninterrupted. By that time the Owatonna branch of Bell had nationwide direct dialing, but there were still nineteen rural telephone companies using operator exchanges for 722 customers doing business out of the Owatonna exchange.

Schroeder Cashway

In the mid-1950s Harold Schroeder, who farmed south of the village, started selling corn cribs, grain bins, wagons, grain boxes, and hoists from his farm. One year he sold seventy wagons, boxes, and hoists. Other dealers protested because he was not listed as a direct machinery dealer. In 1959, to avoid any confrontation, he purchased the village's second schoolhouse, which had served as the barbershop/pool hall and general meeting place since 1922. It became his residence and place of business. At that time he added other short-line products and the business went well. The Hayes Lucas and Botsford Lumber companies merged to become United Building Center, and in the process the Meriden yard was closed in 1963. Schroeder bought the Hayes Lucas facility and moved much of his inventory there, plus he added a full hardware line and bulk milk tanks. Soon after, he became involved with International Minerals and erected a warehouse directly west of the hardware store and sold fertilizer and chemicals under the name of Meriden Agricultural Center. After he got more involved with the grain business the name was changed to Meriden Grain and Farm Center.

In 1965, the National Farmers Organization (NFO) asked Schroederif if he would store grain for them. He built some bins and became a competitor of the Meriden Grain Company, which was owned by the Owatonna Farmers Elevator. It was not long before he was buying two to three million dollars of grain each month in addition to carloads of fertilizer costing $40,000 per load. In 1969, he took in two partners, a move he soon regretted, but in October that year they purchased the Meriden elevator. On

January 7, 1970, misfortune struck when one of his partners drilled a hole in the gas tank of a snowmobile in the elevator office. It exploded and caused a $400,000 fire that "consumed the structure but I continued in the grain business."

In 1973, Schroeder dealt with farmers in hedged-to-arrive contracts and then placed many grain contracts with the terminals. Soybeans jumped from $3 to $12 a bushel, and about twenty-five farmers did not deliver. In 1974, he had a judgment brought against him by the Minnesota Supreme Court for $391,395 in favor of the Victoria Grain Company of Minneapolis. He said, "We almost went broke overnight." He got out of the grain business but continued in the fertilizer, hardware, and short-line machinery business until 1975 when he had a "four-day auction of hardware and machinery," after which the fertilizer business was acquired by Darrel Johnson and George Jones. I was unable to determine how long they were involved.

The Creamery

The Meriden creamery was larger than most small rural cooperatives and had excellent management, so it continued to grow and remain profitable. In late 1959, Uecker was instructed to purchase a roll drier, an evaporator, a larger churn, a pasteurizer, and a separator. That year the creamery had a net profit of $15,869, which was put into equity reserve; $8,000 was paid out in dividends from the 1954 reserve.

At the February 1960 annual meeting the county agent pointed out that hay acres were unchanged as were cow numbers, but production was up. It was proof that farmers were doing a better job, but he warned them to be careful in their use of penicillin so there would be no residue in the milk. The agricultural director of station KDHL spoke on the benefits of artificial insemination and milk quality. The meeting ended with a discussion about going into skim milk and buttermilk drying as a method of expanding. The vote was seventy-seven for, one no, and one abstain for moving ahead.

At a special meeting in March several reports were given about experiences others had had with drying and the problems with enlarging the building if they moved to go ahead. The vote was fifty votes to proceed and eighteen nays. The bylaws were changed to read unlimited indebtedness and the board was directed to appoint an advisory committee to proceed.

In January 1961, the Creamery Cooperative Association, which operated in the leased Borden plant, announced that it had paid over $1 million for butter fat and skim milk, had added 104 patrons making a total of 283, and had produced nearly 1.4 million pounds of butter at a new low cost of $0.0496/pound. In 1965, the Meriden creamery purchased 1.66 million pounds of whole milk and was still expanding rapidly.

Uecker realized that other larger creameries were facing the same challenge that Meriden had, and when he learned that the Borden plant could be purchased he realized that securing it would be a better way to face the future. In 1966, seven creameries formed the Southern Dairy Association and purchased the Borden plant for $2 million. Besides Meriden, the association was made up of Waseca, Otisco, Owatonna, Moland, Dodge Center, and Brownsdale. At that time the Meriden creamery closed and Uecker was made manager of the reorganized cooperative's plant with its forty-two employees. Several other creameries soon joined the movement with their 800 farmer patrons. In November 1969, the Association joined American Milk Producers Inc. (AMPI), which was the nation's largest dairy cooperative.

Kath Enterprises

In 1932, the Albert Kath family moved to Meriden Township where Kath had rented eighty acres. They had several children, and the farm was obviously too small to make a living even in the best of times. I recall that when Edward entered school at Meriden my dad had told me how poor they were but how hard the parents worked. Within a short time the family moved to Deerfield where, within a few years, they were able to purchase a farm. In the years that followed my dad and Albert did considerable business. While growing up, Edward worked as a hired man for five different nearby farmers. After service in WWII he worked at several jobs including working in a café in Waseca where he met and later married Kay Fogarty.

In 1957, a friend, Marvin Ebeling, who farmed in the area and had purchased a store at Dexter, employed Eddie and Kay to operate it. They made a success of that store and wanted to get into their own business. Leo Thompson, who was doing well in Meriden, wanted to sell, and in 1964 Eddie and Kay purchased that business. Thompson had started a little café in the store, which was also successful. It fit in well with what Kay and Eddie wanted to do, although Eddie said, "We sold big hamburgers way too cheap."

The poultry business was a major reason for the success of that store, as farmers bought as much as 100 tons of Big Gain poultry feed per month in addition to groceries. For a period of time they were a leading dealer for Big Gain, and the feed business kept their sons, Jeff and Carley, busy making deliveries. Then, in 1983, a nationwide avian flu epidemic hit and, combined with negative propaganda about eggs and cholesterol, many smaller farmers were forced out of the business. Fortunately, by then Eddie had become involved in the propane business, which was used both for heating and crop drying, and it became their most profitable enterprise. They realized the village businesses were losing customers, but they worked directly with the farmers so their major income was not affected. Their family kept "very involved in the church and in the community and really enjoyed their life there until Eddie and Kay retired."

The Post Office and Drache Truck Line

My Mother, Anna Drache, had been postmaster since 1929 and had never taken any major time off during those years, so she had accumulated a considerable amount of leave time and was encouraged by her superiors to take a vacation. The trucking business was doing well and Dad was in the mood to take more time off than just going fishing. Walter Zatochill was well able to handle the business. On August 18, 1960, to prepare for that leave, Marion Davidson, wife of Myron Davidson, the local Raleigh dealer, started filling in at the office. At the same time, Dick Rietforts had turned sixty-five and he also had not taken much time off since he had purchased the blacksmith shop in 1928. He was anxious to return to Germany to visit relatives. The Draches and the Rietforts, along with their daughter Gladys, took a ten-week trip to Germany in 1961 and had a great time. After they returned I asked my dad how the business did while he was gone and he replied, "It made the most money of any two months ever."

In 1967, the post office reverted back to fourth class because business had declined since the closing of the bank, Hayes Lucas, the creamery, and the sharp decline at the Davis Store with the reduction of the poultry business. These were the four major postal clients. On April 29, 1968, the postal inspector decommissioned Anna Drache as Meriden postmaster, and she took a year of sick leave plus her remaining vacation time. The following day Marion Davidson was made acting postmaster with Paul Drache serving as substitute.

In August 1968, the postal department announced that the Meriden postmaster position was open at a salary of $6,348 a year. Marion Davidson was the only one who took the examination and received the position. The final salary for Anna Drache was $8,064 plus $30 a month for rent and utilities. During the final weekend of October 1969, the post office fixtures were temporarily moved to the bank building because none of the promised equipment had been delivered. In those days postmasters in the small offices had to purchase the equipment, which Drache did in 1928 when she was notified that she would be named postmaster. That equipment had served Samuel Grandprey, Henry Luhman, and Anna Drache since 1904, and when she retired it reverted to her and was then donated to the Village of Yesteryear.

In early December 1967 Curtis Yule, of Medford, came to Meriden to see Paul Drache about buying his trucking business. Yule visited with all the Drache customers, and on December 29, 1967, after examining the books, he purchased the business on a contract and made a down payment. On January 31, 1968, the Railroad and Warehouse Commission approved the transfer of the rights. Paul Drache's business career in Meriden that started in 1917 came to an end.

Drache started employing drivers within a few months after he purchased his first truck in 1929. By the late 1940s the truck line employed fifteen drivers, but most of them were from outside of the township. By the time he sold the business, over eighty drivers and eleven women, who worked as domestics or secretaries, had been employed by the Draches. In December 1967, Yule took some trucks and two trailers to Medford, but two trucks were housed in Meriden through 1971. Walter Zatochill remained with Yule for a few months to help with the transition while Walter Voss, another Drache driver, drove for Yule until he retired in 1977.

Palas Garage

In 1917, Louis C. Palas announced that he was going to build a garage north of the tracks and east of the county road. In 1919, he constructed a 40 x 70 foot building, which was the first garage in the village. Palas built a very modest tower that generated electricity into batteries. This served the business for many years. Some time during the late 1920s, his son, Harvey, took over the business and remained there until he closed it in the 1970s. Harvey purchased old cars and removed the body, then converted the chassis to two-wheel or four-wheel trailers. The garage had one gas pump and a few regular customers, but Harvey was kept occupied by people who had "special projects" and did not have the needed skill, tools, or time to do the work.

I visited with Harvey on my walking trips around the village and always enjoyed the experience. Most of the people living there during those years probably recall him sitting on the bench in front of the garage watching the traffic. In his early years he enjoyed speedy cars. In 1934, he had a new Ford V8 coupe equipped with police gears. His antics with that car caused frequent conversations. He died in 1985, and the building has since been razed.

Davis Grocery

Diemer Davis was active in the Southern Minnesota Associated Grocers Association and was also very involved in the community. In 1946 the Association had 151 grocers in ten counties, and in 1974 it had the same number of grocers in seventeen counties, which is indicative of what was happening in the rural areas. The Davis store business continued to be successful, but after Diemer's death it proved too much for Lil Davis, so she decided to sell. When she was unable to sell the business, she liquidated the inventory and, on April 15, 1973, sold the building to Jerry Kath who converted it into two apartments.

Blacksmith Shop

It was mentioned that Dick and Gertrude Rietforts took a ten-week vacation in 1961. He was sixty-five and the trip signaled his intention to reduce his activity in the blacksmith business. In 1985, he recalled the technical evolution in the business from repairing machinery that still had wooden parts, then the fading away of horse shoeing, followed by the loss of sharpening plow shares when throw-away shares were introduced. Next electric welders became available and that work was gone. Then farmers built their own shops, so there was little left for blacksmiths to do.

Rietforts said the most dramatic change came when farmers stopped delivering their own cream and shifted to whole milk hauled to the creamery by a few trucks. "Then the farmers raced to town on the better roads and had no reason to come to our little town. At first I got my supplies shipped in by the railroad. Then, in the 1930s when Drache started hauling to the Twin Cities, he picked up my orders and delivered them to the shop. But when he sold in 1968 I had to buy the steel products from McNeilus Steel in Dodge Center." He made a parting comment about the rent that the township charged for his use of the building. When he started in 1928, it was $8.50 a month. Sometime after WWII it was raised to $10, and in his final years he paid $15. The town board was well aware of his worth to all the farmers and others in the community. As business dwindled he continued to work part time until 1973 when he closed the shop for good. Gertrude Rietforts, who kept the books and saw him off to work every day, commented when the building was moved to Farm America in 1981, "I never thought he was going to get to be so old the way he worked. This much I'll say. He was never crabby when he came home."

Shipping Association

Kenneth Brase became the trucker-manager for the Meriden Cooperative Livestock Shipping Association in 1955 and was still in that position in 1970. In January the group held their annual meeting in the VFW hall in Waseca. Robert Armstrong was elected president; Milan Prechel, vice-president; Leonard Scholssin, secretary/treasurer; with Norbert Abbe and Joe McShane, directors. Once the stockyard closed in 1929, the shipping association had little impact on the village other than the fact that most of the members were from the Meriden trade territory.

By the 1970s, the population of the village had declined to 155 who were either retirees or worked in Owatonna or Waseca. Except for the Kath store, the Renchin tavern, and a garage, there were no accommodations for daily living available. Paul and Anna Drache left for Arizona each fall, and the Grandpreys spent much of their time with friends in Owatonna. Both couples agreed that life in the village had become "very dull."

Agriculture

Dairy, Hogs, and Poultry

The 1959 agricultural census indicated that the county had 1,715 farms of which 1,602 were commercial farms and 1,400 were dairy farms that averaged 212 acres and 14 cows per farm. They produced an average of 8,600 pounds of milk annually in contrast to the state average of 7,640 pounds. The county was still one of the top milk producing counties in Minnesota, which was still the number one butter producing state. The Meriden DHIA had 23 herds with 598 cows, for a 27 cow average. Those herds averaged 11,343 pounds of milk and 445.6 pounds of butterfat. The four top herds averaged over 500 pounds of butterfat and the low herd now had a higher production average than the high herd did when the association started. The Minnesota average for 1959 was 267 pounds of butterfat, proof that the local DHIA had done its job well.

By 1972, the number of cows in the county had declined to 13,928, but their average production had increased to 10,800 pounds, still far above the state average. But the DHIA herds had increased to over 14,000 pounds. The cows in the Meriden association earned a return of $223.68 over feed cost and produced an average return of $219 per cow greater than the non-DHIA herds. At the 1972 annual meeting, W. P. Jones received a special award for being the instigator and one of the founders of the Meriden DHIA.

The dairy industry continued toward larger and more efficient dairy farms as the number of herds in the county declined to "less than 800" following the drastic national decline. Predictions were made that in 10 to 15 years the average milk production per cow would be 20,000 pounds. Milk production held firm, but the number of creameries in Minnesota fell from 600 to 363 while the number of plants that produced dry milk products for the food industry grew. All the county creameries except Hope were closed.

Many township farmers had raised hogs and poultry along with dairy. Hogs were so important to the producer that when cholera entered any area there was fear among the farmers. That plague was difficult to control, and it was not until 1972 that Minnesota was officially recognized as hog cholera free by the National Hog Cholera Eradication Program and became the 38th state to achieve that status. Smoked pork products were a key to the diet as lard was to the cook. Lard was used to make soap and provided a way to preserve food in large Red Wing crocks. Lard had many uses, but as other cooking oils were introduced it became a burden to the packer and over-fat hogs were docked. From 1958 to 1967 the amount of lard per marketed hog was decreased by six pounds, which increased the income of both the packer and the producer.

In 1961, Nebraska farmer Paul Bird came to the township and purchased 200 acres. He grew corn, soybeans, and hogs and built what was probably the first slatted floor hog barn in the township and the county. In 1968 he was named Steele County Young Farmer.

The 1959 census indicated that Steele County had the largest concentration of chickens per square mile in Minnesota and the highest rate of layers in the state. Flocks of 300 to 500 birds had experienced the biggest drop and were replaced by flocks of 2,000 to 6,000 birds. Although not a township farmer, Rodney Young, of the Havana and Owatonna area, was an example of what took place in that sector of agriculture. He was one of fourteen farmers nationally to qualify for the Ford Almanac because he had adequate records to prove what could be done with poultry. His 10,000-bird flock produced 259 eggs per hen per year. A dozen eggs were produced with 4.2 pounds of feed at a cost of $0.21.

Hy-Line Poultry Farms in Owatonna had twenty-four full-time and twelve part-time employees to hatch 570,000 chicks every three weeks. In 1966 about 700,000 pullets that produced about $4.2 million worth of eggs were placed in the area. In 1969, the Grain Terminal Association erected a feed processing plant along the Minnesota, Dakota, and Eastern tracks west of Owatonna to provide prepared rations for those farms..

Mechanical Changes

Some of the mechanical changes that took place during the period of this chapter may seem insignificant to the reader but each of them helped the farmer do a more efficient job with less human effort. In June 1960, a new PTO-driven grinder/mixer came to the market. As agriculture became more capital intense, the concept of leasing equipment was developed and spread rapidly among the larger farmers. By July 1960, the first ad for leasing combines appeared in the local papers. As farmers dropped their livestock enterprises and concentrated more on cash cropping they traded the ear corn picker for a combine, which increased the need for on-the-farm drying. In 1960, OMC produced a cylindrical drier that sold for $4,000. For those who maintained their livestock enterprises, the silo unloader became available from Van Dale. The auger feeder distributed down the bunk line. The skid-steer self-propelled loader had been developed in the 1950s, and in 1964 OMC displayed its first model, which was rapidly adopted in agriculture and industry. Only three years later OMC introduced the four-wheeled self-propelled loader. The first commercial barn cleaners were developed in the latter 1930s but it was not until the 1960s that a good all-weather cleaner was available. In 1966, the Ost dealership in Owatonna sold its largest tractor, a D-21 Allis Chalmers that pulled an eight-bottom plow, for $14,00. This county farmer had 1,740 acres, of which 1,200 acres was in corn, and he raised 140,000 turkeys and 3,000 mink, in addition to beef cattle and hogs. By then the four-wheel-drive tractor was well proven, and it was only a matter of time before one would be on a farm in the township.

In 1968, OMC introduced a four-row stalk shredder for corn stalks because the sturdier corn plant caused problems when plowing and slowed down drying in the spring. Farmers became more sensitive to timely operations and shredding became a must. In 1970, a heavy duty baler was marketed that threw the bale in the wagon, which eliminated the back-breaking job of stacking the bales on a flat-bed wagon. In 1973, OMC introduced a self-propelled mower-windrower-uni-stacker, which accumulated the hay from the windrow, chopped, and blew it into the stack-shaped box, which formed it into a stack of six to eight tons and then self unloaded it.

In 1968, over 50 percent of the corn crop was harvested as shell corn. To speed up harvest the Mayrath Company developed six-inch grain augers from thirty-eight to fifty-four feet in length and an eight-inch model that was from thirty-seven to seventy feet in length. Corn was stored in round bins and replaced the corn crib, which had been a long-time fixture on corn-belt farms.

Tiling was a real drudgery in the days when it was done by hand digging the trench and then laying the clay or cement tile by hand. Tractor-drawn trenchers were introduced in the 1940s, but there was still a considerable amount of manual labor involved. A big break through in tiling came when plastic drain tile was introduced. Tile could be laid by machines that relied on lasers to control the depth, and the rate of laying the line was determined by the power unit.

In the boom period for agriculture in the early 1970s, the county had eleven implement dealers, seven of which were located in Owatonna, three in Blooming Prairie, and one in Ellendale. In 1974, the dealers all announced that they would not be displaying any machinery at the county fair because "there was a nation wide shortage of equipment. Machinery deliveries were running a year behind." The canning company reported that "Because of the labor shortage they may leave vegetables unprocessed because there was a shortage of willing hands to process this summer's vegetable harvest. . . . Fewer and fewer high school and college students are willing to put in the long hours that work in the fields and canning factory demands."

Biotechnology

The advantages of soil testing and commercial fertilizer were known, but beyond applying manure from the livestock and plowing down a green manure crop, most farmers did little to enhance their soil. That changed, and the decade of the 1950s became the decade of soil testing and the adoption of commercial fertilizer. The cost per acre rose but so did the production per acre. This production process was perfected again in the 1960s when chemicals were introduced to aid the crop in its fight against insects and weeds.

Hybrid seed corn was well adopted in the area after WWII. Then the corn seed was modified to provide special needs. In 1962, a new silo blend 95-105-day maturity seed, which was capable of increased tonnage over earlier blends, was available at $3.90 to $4.90 a bushel. Hybrid crops were sturdier than the crops of the past, and farmers soon learned that they were capable of being "pushed" to produce more.

In 1964, the county agent noted that herbicides applied to crops early in May gave "very good control in conquering weeds." He recommended atrazine for corn and trifluran on soybeans. In a column in 1966, the agent urged farmers to use parathion and diazinon on plants that were to be used as forage. They were both recommended to fight corn root worm beetles as well as the western and northern beetles. The Gandrud Company, of Owatonna, had its own demonstration farm, and in 1966 it had plots where it used both pre-emergence and post-emergence chemicals such as Banvel-d, granular Randox T, atrazine, 2-4D amine, and others to illustrate the options that the grower had. That summer a Canadian thistle demonstration plot on the Steinberg farm in section twenty-four in Meriden Township used seven different chemicals to show how each worked on that plague to agriculture.

Every year farmers observed that new seeds, fertilizers, and chemicals were helping yields, and the extension service, along with innovative farmers, had learned that timeliness was essential to maximize production. When the soil temperature reached fifty degrees it was time to be in the field because "early planting resulted in higher yields, there was more sunlight utilization, better pollination, shorter plants, and corn was more likely to reach maturity." Post-emergent weed control was most effective in early season.

Historically, there have always been critics of change, and adopting commercial fertilizers and chemicals to food production were no exception. In December 1970, the Gandrud Company sponsored a meeting of a biochemist from Rutgers University and the local Association of University Women. The expert stated: "Proper use of pesticides is necessary if we are to feed our present population now and in the future. . . .No research evidence shows significant injury to humans or the environment where pesticides have been properly used." Russ Gute, the county agent, pointed out how many died in car accidents, falls, and from shooting, but only 200 from all pesticides, "which is less than the deaths caused by aspirins. Yields of all crops have increased and were likely to increase."

From 1960 to 1971, the number of farms nationally with sales of over $40,000 had increased from 113,000 to 253,000, while farms under $20,000 in sales had dropped sharply. Then, in December 1972, the huge wheat sale to Russia resulted in higher prices and the decade of the 1970s proved to be the most prosperous on record for American agricultural. A 1973 study by the USDA indicated that one Midwest farmer could raise corn and soybeans on an 800-acre farm with sufficient capital and mechanization. Such an operation required a $700,000 investment. The average size farm had increased from 140 acres in 1940 to 385 acres in 1974. This forecasted that greater change was ahead. An editorial in the *Photo News* cited the *Christian Science Monitor*: "By 1980 the developed world would be virtually consuming all the food it could produce. Food production must double during the next generation."

The declining number of farmers in the county caused a change in commissioner districts. District No.1 in which Meriden was situated was enlarged to include Berlin, Deerfield, Lemond, Medford, and Summit and the villages of Ellendale and Medford. The basic industry in the township was producing more that ever, but it was losing population and the village businesses were dying. This was happening throughout the nation.

Crane Creek

In 1963, Ewald Wilker, Gene Phillips, and Eldon Neuman represented Meriden and Deerfield on the 66,000-acre Crane Creek Watershed Flood Control committee. That year the Crane Creek project was given its final okay of $1.49 million under the Small Watershed Act of 1956. The watershed project was expanded to include flood control, development of wildlife areas, and drainage outlets. It was called Judicial Ditch #24 and extended twenty-six miles from near the village of Clinton Falls to Rice Lake and the Moonan marshland northeast of Waseca. By the time it was finished the area encompassed 77,000 acres and had a total of fifty-eight dugouts for loafing and nesting sites for ducks in the Waseca marshland area.

A related news item reported that Harry Andrews had constructed a sod waterway on his farm in 1945, the year before the County Soil Conservation District was formed. It is believed that he was the first farmer in the county to construct such a waterway. Andrews reported that their farm had a fifteen-foot drop to Crane Creek and the waterway was necessary to stop the intense erosion.

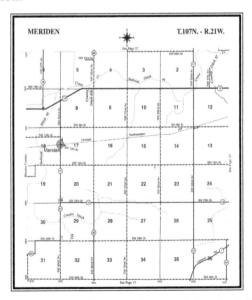

This plat is of the Crane Creek Watershed Project. A GIS – Steele County image.

A Dynamic Agriculture Conflicts
with the New Direction for Agriculture

Vernon Ruttan, Chairman of the Department of Agricultural Economics at the University of Minnesota, wrote in 1965, "If the techniques used on all farms producing over $40,000 gross sales were applied to all agriculture, only 400,000 farms could meet the nation's needs." Ruttan knew that the average output of an agricultural laborer in India was about 1/50 of that of an American farm laborer, and in very few countries was the output one-fifth of the American worker. He explained, "The process of growth in agriculture, in a progressive economy with rising real wage rates, is one of education, technology, capital, and shifting surplus workers to other production which enables total growth."

By 1964, farm numbers in the county dropped to 1,569 but gross farm income rose to $17,841,805 of which $1,678,868 was spent in diversion and feed grain payments. Census data reported that per capita income for those living on farms was $4,753 in contrast to $4,330 for the nation. However, in 1965 the Agricultural Stabilization and Conservation Service (ASCS) held three auctions to dispose of the bins the government had to store surplus grain. The storage program was no longer part of the total program.

Some of our farmers were blazing new trails in agriculture that were benefiting all of society, but many farmers were not happy with the conditions. In 1955, the National Farm Organization (NFO) was founded to protest low prices and government policies. In September 1964, the organization had reached the county, and action by members "led to five county incidents that all resulted in damage." One of those incidents took place on U.S. 14 within the township in which a sickle bar was placed on the highway and caused damage to a truck loaded with cattle. In January 1968, the county organization joined a thirty-state holding action on small grains "because prices were lower than twenty years earlier."

The government was well aware of the problem and wanted to reduce the feed grain surplus by increasing the diversion from 10 million acres to 30 million to improve prices. Farmers were asked to place 20 percent of their corn base into the conservation program. In March 1969, the *Photo News* devoted five pages to agriculture; one was to government programs, one to the NFO and collective bargaining, another to the Farmers Union on the irony of surplus and hunger, a fourth to the Farm Bureau and too many farmers and supply and demand, and the final page was entitled "Is Corporate Farming a Problem?" The NFO and the Farmers Union said yes, and Farm Bureau said no.

When the Food Stamp Act was passed in 1964, Steele was one of 235 counties to be added to the program. President Johnson said, "The purpose of the program is to give low-income families more food purchasing power for improved nutrition and also to channel the farm surpluses of the country toward low-income families." What was not said was that the cost of the program would be charged to the USDA. As stated above, the picture in agriculture changed sharply in December 1972 after the Russian grain purchase. Prices hit new highs, and by June 1973 the government banned the exportation of soybeans to Japan, which needed them for protein for the people. The dilemma in agriculture continued.

What Happened to the Schools?

The village school had switched to an oil furnace in the 1930s, which was a big help for the teachers who were expected to do the janitorial work. Probably more important it meant that the school would be warm when everyone arrived in the morning. In 1960, Lorraine Bock was employed to do all the janitorial duties. That year the Rural School building at the county fair was changed to the Fine Arts building and would feature the work of amateur and professional artists of the area. In 1963, the seventh and eighth graders were taken by bus to Owatonna leaving the school with six grades. As of 1960, only Hope and Meriden of the forty-six remaining rural schools had two teachers to teach their 755 students. A letter to the editor in the *Photo News* questioned why we persisted in having "One Room, Back Shanty Schools? . . .Why do we continue to have Horse and Buggy Schools?"

In 1967, the county still had thirty-one operating rural schools with 483 students. The state ordered that all rural schools should be closed by 1971. The Owatonna school district had to add eighteen more classrooms to handle the influx, and the school system had to make provisions for ninety children in the Head Start program. All the rural schools were closed as of June 1971. The final adieu was carried in the August 20, 1971, issue of the *Photo News*. It read: "The Little Red School House, the place where many millions of Americans past and present went to get an education, has been a hallowed institution, a heritage looming as big as the revolutionary signing of the Founding Fathers. Today the majority of Americans are urban dwellers, and the demise of the Little Red School House is a product of the population shift."

A Hodge Podge of Daily Events

By 1960, the Steele-Waseca REA had grown to 4,750 members and by 1965 it had 5,500 members. The number of farms kept declining, but rural life became more attractive as such amenities as telephone, lights, good roads, fire protection, and school buses were available. Many preferred to live in the country and commute to work in the city. The cooperative had twenty-six employees of whom eleven were so stationed with their service trucks that no member was more than thirty-five miles distant. By that date each customer was using about 650 kilowatt hours per month.

Owatonna provided space in its fire hall for the rural truck that was operated by the city fire department. To receive fire protection, each rural non-farm residence was assessed $150 plus $30 per hour for the fire truck. Meriden had the benefit of the large creamery well, which was kept in operation to supply water for fire protection.

In 1962, a county-wide bookmobile service was re-established to give better service to the rural areas. Each Thursday the book van stopped at Steele Center, Hope, and District 36 schoolhouse in the southeast part of the township. Next, it traveled to Pontoppidan Church and made its final stop for an hour at the Thompson store in the village.

Steele County had twenty-six 4-H clubs in 1960 with a total membership of 735. In February that year, Bernice Scholljegerdes, member of the Meriden Meadowlarks, won the county 4-H Radio Speaking contest. Her topic was, "Why I Am Concerned With the World Refugee Problem." That year she was also named Steele County Dairy Princess and entered the state Princess Kay of the Milky Way contest.

That year Kenneth Dinse was elected president of the Owatonna FFA chapter. He was also a member of the county 4-H livestock judging team along with Bill Kriesel and Donald and James Gute. They attended the International Livestock Show in Chicago to compete in the National Judging contest.

Another 4-H club was added in 1962, making twenty-seven in the county, and the membership increased to 832 members. Dairying was still the most popular project with 351 participants, followed by swine with 323, food preparation had 285 enrolled, and 226 had a clothing project; others were involved with sheep, beef, and shop.

After serving as organist for fifty years at St. Paul's Lutheran Church, Paul Drache decided it was time to retire. At the Christmas Eve service that year he received a plaque.

Rural postal patrons were reminded to mow the weeds in the ditch around their mail boxes because "where the weeds were left it was more likely to catch snow and block the road. Keep the approach to your mail box free of obstructions which will not permit serving the box without leaving the car. Rural mail carriers are not required to perform service on foot when roads cannot be traveled with conveyances ordinarily used by them."

Paul Steinberg, a member of the Meriden Meadowlarks, had the grand champion Ayrshire cow at the state fair and was presented the top award at the 4-H dairy banquet. Later in 1964, Richard Steinberg and Mark Buecksler, both of the same club, took top honors at the Blooming Prairie Black and White show judging contest. Steinberg was the only contestant out of over 100 to earn a perfect score.

On December 16, 1969, a "wild man" entered the Davis store and assaulted clerk Benita Volkmann and then knocked down Orvetta Schroeder and broke her hearing aid. Soon after the event the county sheriff nabbed him. That was the major crime in the township during these years.

By 1970, the twenty-seven 4-H clubs in the county had 923 members. The Meriden clubs were particularly active during this period. The Meadowlarks had Mr. and Mrs. Arnold Abbe, Mr. and Mrs. Robert Armstrong, and Mrs. Lorenz Wilker as leaders. The Meriden Skippers were aided by Mr. and Mrs. Emil Steinberg, Mr. and Mrs. Robert Reul, and Marvin Buecksler. Norman Abbe was a delegate to the 4-H Dairy Conference in Chicago. By the 1970s the extension service nationally was receiving far more calls from urbanites than they were from farmers. That trend was reflected in an increased interest in garden clubs and urban interest in 4-H.

In 1974, the township led the county in raising money for the United Way of Steele County by raising $959. Owatonna Township was second with $865. Social life was changing, partly because more residents from the village and from farms were working in Owatonna. Some people from the township began taking trips south for a month or longer each winter, and others still had card parties or people over for an evening meal. With the closing of the rural schools more of the rural youth had contact with people from Owatonna and stretched their social life beyond the limits of the township. Except for a few farmers living in the eastern sections of the township, very few township farmers played a role in county-wide activities in early decades. The trend changed in 1967 when James Andrews and Odell Knuston were named superintendent and assistant, respectively, in the sheep department at the county fair. Paul Schroeder and Carl Oeltjenbrauns were superintendent and assistant, respectively, in the grain and grass department.

Changes That Impacted Life in the Township

The director of the Owatonna Chamber of Commerce, who had resigned to take a position in a much larger city, wrote a farewell address that included this comment: "In a time when many small agricultural communities are drying up and withering on the vine, Owatonna is headed in the other direction. . . . This is because of its fine industries and outstanding owners." A month later one of the county rural churches had a priest from the Catholic Rural Life Conference speak on "The Threat to the Small Family Farm." In the previous topic the mechanical and biotech changes that were taking place in agriculture were adopted by progressive farmers to improve the lives of those who tilled the soil. Fewer farmers would be needed to meet production demands. The members of the Owatonna Gavel Club had realized early what was happening in the farming area of the county and knew they would have workers from the county farms. By the 1960s there were 472 firms in the city that relied on them.

An article on the subject read: "The conservative figure [long-range population] is based on the assumption that part of the impact of economic growth will be absorbed by population growth outside of the city limits, such as fringe areas and other small towns and urban centers within commuting distance as had occurred in the last three years." It was pointed out that manufacturing alone had increased its employment from 1,200 in 1950 to 2,200 in 1963. The industrial growth on the west side of I-35 that ran through that area showed a growth in assessed valuation of 60 percent residential, 30 percent commercial, and 10 percent industrial, which was deemed a good balance. The thirty-one manufacturing firms had shown a 60.2 percent growth, the greatest of any other sectors, followed by trade, which had 1,120 employees, and finance with 518. The service sector had 68.1 percent female employees and finance had 61.4 percent. In 1960, 29.1 percent of the wives in the county were working outside the home. By 1963, that figure had increased to 38.1 percent. The reasons given were labor-saving devices and prepared foods. They provided 38 percent of the family income.

Changes took place in Owatonna to provide for its growing population. The Town and Country Plaza, a shopping center that had six large and many smaller stores, became the shopping center for that part of town. Downtown the post office was razed to make space for an expansion of Federated Mutual Insurance. Hotel Owatonna, a city landmark, was razed to make room for retail stores and gain parking space. An accompanying editorial to the news article stated: "The nation is becoming more mobile and fluid . . . the home town merchant can no longer count on 100 percent loyalty and patronage by buyers from his home community." The better highways made it possible for Rochester, Austin, Albert Lea, Mankato, and Faribault merchants to compete. A follow-up article, "Small Towns, Rural Areas in a Tax Bind," stated that there was too much reliance on property taxes, which were needed to support the schools. But as soon as the students graduated, they took their increased earning capacity and tax paying ability to the cities, which then reaped what the rural folks had sown; this was cited as one of the causes of the rural hardship.

While rural people were worried about their declining life style, the urbanites complained about inflation and the rising cost of food. An editorial in the *Farm Journal* defended the farmers: "We suggest that the Secretary of Agriculture, the President, the rest of his administration and Congress concern themselves more with the real cause of inflation: government spending and expanding fiscal policies. Let's have less wailing about food prices and some real shouting about that." At the same time two Minnesota politicians said, "We needed a new strategy to revitalize rural America—we must stop the migration to the cities." Another said, "The highest levels of rural poverty were among those people who were not farming." Then, in 1972, the president went to China and the headline read, "The week that changed the world. . . . Thereby the balance of power in the world has changed." That became a major step in globalization, which would have a major impact on all society.

While highways were being constructed and urban centers were growing, the world was shrinking because of globalization, and railroad passenger service kept declining. On July 23, 1963, the last of the "400" CNWR passenger trains stopped at Owatonna. It signified the end of all east-west passenger traffic on that road. In 1964, the CNWR petitioned to cut station agents at Claremont, Janesville, and Kasson and remove the station buildings.

A 1966 *Photo News* article said it well: "The underlying causes of the public's desertion of this service are well known." The Chicago, Rock Island Road petitioned to discontinue passenger service to Owatonna on the train that ran between the Twin Cities and St. Louis. Operating the Owatonna station had resulted in a loss of $209,545 in 1965. However, the north/south service on the Chicago, Rock Island, & Pacific Roadroad continued on the route between the Twin Cities and Kansas City, but in June 1969 they asked for permission to shut down that service. Passenger traffic had been declining, but the loss of the first class mail service assured an operating loss. As late as 1954, Owatonna had twenty-four passenger trains daily, but on July 18, 1969, the last passenger train passed through town.

While passenger service was being discontinued, freight service was having a revival after suffering heavy losses to the trucking industry on short hauls and less-than-carload business. By 1965 covered hopper cars offered easy loading and unloading and could haul a 100-ton of grain instead of the previous fifty-four-ton box car that had to be loaded through side doors and required the use of costly grain doors. Tank cars were increased from 8,000 to 30,000 gallons. But even more dramatically was the introduction of 20,000 elongated flat cars that could piggyback two truck trailers. That took 2 million truck hauls off the highways.

In 1960, Owatonna already had twenty-three firms that marketed their products world wide. In 1964, the State School reduced its farming operation and had 160 acres for sale located south of #14 and west of what would become I-35 (NE of section 8, Owatonna Township). The city was interested in that land, which the state suggested would probably be $250 to $500 an acre. In was laid out to be an industrial park, and in 1966 Wenger Corporation was one of the first companies to build there.

The rapid industrial expansion caused a continued housing shortage in the city, and in 1969 the first factory-built homes were approved. In October, ADM Industries, Inc., manufactures of products for the mobile home industry in Iowa, Minnesota, South Dakota, and Wisconsin, moved into town. In March 1970, it was announced that a 300-unit low-income housing facility was to be built on the north side. In April, OTC acquired three companies that made plastic products and window hardware for homes. OTC became a major supplier to the window industry. It attracted other manufacturers and in October 1971 the DeRose Industries, the tenth-ranking manufacturer of mobile and semi-mobile homes, established a plant that it expanded almost simultaneously to meet the demand for its units.

Jostens reported that they had experienced a steady labor shortage in the past decade and had recruited actively throughout southern Minnesota. They did not think that the housing shortage was the total problem—"Owatonna just did not have a large enough population." In 1974, after years of deliberation the voters passed a referendum by a vote of 2,500 to 180 to buy the seventy-five-acre West Hills site with twenty-seven buildings for $200,000.

The city had good reason to be optimistic, for from 1961 to 1974 the labor force grew from 5,782 to 8,662. In the later years several thousand people commuted every day to their workplaces. The members of the Gavel Club had foreseen that the declining need for workers on area farms would make them available to fill the rising demand for help in Owatonna.

First State Bank Minutes; Creamery Association Minutes; Arthur H. Uecker, cassette tape, 20 July 1999; Dale Lebahn, telephone interview, 20 June 2010; *Waseca Herald*, 16 December 1961; Meriden RTC; Harold Schroeder, Recollections; Edward Kath, interviews; Drache, *Creating Abundance* 153; Drache, *Tomorrow's Harvest* xi, 27, 264; Vernon W. Ruttan, personal interview, 20 April 1970; Paul Drache, Day Books; Curtis Yule, telephone interview, 20 May 2009; Dick Rietforts, personal interview, 21 October 1985; Peggy Kennon, cassette tape, 8 January 1986; *Encyclopedia of American Agriculture* 255; Chronological Landmarks 80; History of Meriden School District 48/2118; Severson 137.

CHAPTER IX
Saved By the Siding–1975 and After

A New Era

With most of the businesses closed, except for a limited number of fertilizer, grain, or propane trucks, the village streets were silent for much of the day. The retirees who had cars made a trip to church, went out to see how the kids on the farm were doing, went to the lake to fish, went shopping, or walked to the post office and hoped they might see someone for a brief chat.

A late 1970s picture. L-R showing a former store building now an apartment, the bank now the post office, probably the first picture of a street light, the pool hall/barbershop now an office for a machinery dealer, a garage building now a private storage, the Kath store with the former café now an office for a propane dealer, the first store in the village with an upper floor that had served as an hotel and then as an apartment for many, the once busy store now selling only convenience items, but with the last two gasoline pumps, and storage sheds that replaced a livery barn that had served as the home and studio of a once nationally renowned artist. Sandy Dinse photo.

Kay and Eddie Kath and their children had worked hard in the store since they purchased it in 1964, and because much of the time they had lived on the second floor the business was virtually a 24-7 operation. They gained 599 long-time accounts but after most of the other businesses that relied on daily customers to survive had closed, the flow to their store dwindled each passing week. The first victim in their enterprise was the café, which had done a thriving business as long as there was a steady business trade in the village. Then, in 1976, they closed the grocery store because it was difficult to get regular delivery and to keep fruit, vegetables, and meat fresh because the trade had declined so much. The store became the business office for the feed and propane business, and the former grocery shelves were stocked with an eclectic offering of goods such as candy, light bulbs, cigarettes, a freezer for milk, meat, pizza, and pop, and because they operated several trucks in addition to their own cars they maintained gas pumps so there was gas available for the public. The only place in the village that catered to the public was Renchin's bar, pool hall, and short-order café. Kay and Eddie built a new home in the east end of the village and kept their "store" open during the day. By this date it was known that small towns lost a business for each four to seven farms that disappeared in their trade area, and Meriden was no exception.

The May 31, 1976, issue of *St. Paul Pioneer Press* wrote about Lloyd Grandprey, who had played "Taps" on his bugle since he had been in WWI and after returning to Meriden in 1925 had played for more than 1,000 funerals in the county. The article continued that he even admitted that in recent years he had "squeaked" a couple times and felt blue about it, but he also thought it was his duty to say yes when asked. On September 10, 1977, the Grandprey home caught fire when struck by lightning and a pioneer village couple who had served the community well died of smoke inhalation. Their bodies were found just inside the doorway. They were buried with full military honors.

The other news for 1977 was that the street from the Kath store to the east end was given a new coat of tar, and a few days later there was "trouble" at Renchin's bar. After April 1975, social news was limited because the *Photo News* went out of business. It was no longer profitable because they were not able to secure advertisers for a basically rural-oriented weekly paper. The only Meriden news was that the Rural Health Team mobile unit would be at St. Paul's church for three days. This service was sponsored by Dodge, Steele, and Waseca counties for the purpose of providing free service to everyone for vision, hearing, blood pressure, urine sample, and blood tests.

In May 1980, Tim Arlt became county extension agent and remained in that position until December 2003. Arlt related what was taking place in the township at this period when the village was on the decline. Nationally after the sharp downturn in the agricultural economy in 1980-81, farmers recovered rapidly by adapting energy, capital, management, facilities, computers, no-till, and trace elements into their business and did quite well. According to Arlt the township had a core of very progressive farmers. They had test plots of their own, but they also watched the extension plots, especially those for corn and soybeans. He said that the Meriden leaders were good about adapting to better housing for livestock and kept up very well in adopting new technology. They had a good pace about watching for what was ahead, "and actually pushed us in extension. The early adopters, especially in hog production, were extremely competitive." Arlt related that they were good record keepers, which made them the most profitable because they were quick to adapt when they saw an enterprise was not making money. He stated that it was then that he experienced working with his first "large farmer," Cecil Smith, who farmed an 1,555 acres. In 1989, there were 114 land owners and they farmed average of about 195 acres in contrast to the state average of 295. The largest landowner was John Steinberg, who owned 672 acres.

About 45 percent of the farm families had one or more persons working off the farm to fill the needs of Owatonna industry, which by that date required 6,000 commuters each day. In 1985, Steele County had become a manufacturing-dependent county and agriculture produced less than 20 percent of the total labor income, but farmers in those counties had a higher per capita income than the national average. The USDA no longer keeps records of farms by townships, but Kenneth Dinse and James Andrews of the town board determined that in 2010 there were fifty-one operating farms in the township and, twenty of the farmers lived in other townships. The reader should be aware that the official census definition of a farm is any unit that produces $1,000 income from farm products in a year.

In 1983, the village lost a building that had housed one of its most vital businesses but also had been the most popular community event center when the town hall/blacksmith shop was moved to Farm America west of Waseca. There had been other blacksmith buildings in the village, but they were victims of fire. Rietforts had a busy and rewarding career and had served the community well, so it was fitting that the building with all of the machines and tools that Rietforts required were made available for futures-generations to see.

A Vacant Siding

Dan Peterson grew up in a Lutheran parsonage in New Richland and after his education was employed in the grain trade and enjoyed it.

He somehow got involved in the bird seed trade. He purchased the seed, which consisted of safflower, black oil sunflowers, millet, and milo. The sunflower is the first choice of birds with millet a second choice. His major source was in South Dakota, and his customer was KT Products, a bird seed processor located in Chilton, Wisconsin. Initially he handled the business out of his home in Waseca. Over a period of time he developed good contacts to secure the seed, and KT Products continued to take all he could provide. He started with three trucks and then expanded to five, but operating between central Dakota and eastern Wisconsin without a backhaul proved unprofitable. His big break came when he was able to contract with beer distributors in Dakota that provided him with loads both ways. But the difficulty getting trucks lined up to keep a steady flow of seed on hand meant that he had to find a place with storage adjacent to a rail line.

The vacant siding at Meriden was exactly what he needed, and in 1985 he founded Peterson Grain and Brokerage Company and purchased what had been the Victoria Grain elevator on the east. Then, he acquired the Land-O-Lakes fertilizer plant on the west, Allen Terpstra's 95,000-bushel bin, and five additional 10,000-bushel bins and a large Quonset from Victoria Grain. Peterson added a large scale and five 55,000-bushel bins.

A September 2000 ground view looking east from the junction of Mill Street and the county road showing the Peterson Grain complex that extends east to where the stockyards once stood.

The Dakota, Minnesota, & Eastern Railroad (DM&ER) people were willing to accommodate him. The only expense was replacing the ties for the siding. It could hold seventeen rail cars, but only seven could be effectively loaded on it. In addition he acquired a large milling facility in Waseca, which gave him a combined capacity of several million bushels. (Contrast this volume to the original tower elevator in the village, which had a capacity of 25,000 bushels.) In 1986, the rail line received a substantial rise in status when the Canadian Pacific paid $1.48 billion for the DM&ER and announced that it would spend $300 million to upgrade the system. Soon after they had the facility in order, township farmers called on Peterson him to buy their corn and soybeans, so he became a federal warehouse, "which proved very profitable as long as the product stayed in good condition."

The owners of KT wanted to expand their business and decided they also wanted to acquire their own products. Peterson commented that being in the brokerage business for them was a good business without any major investment, but suddenly in 1992 that changed, and he had to make a major investment in equipment to do cleaning, processing, and packaging on his own. Then he contracted with country elevators. His largest supplier in was the Oahee Grain Company at Onida, which at that time was South Dakota's largest independent elevator and was located on the DM&ER. The Waseca storage gives Peterson the ability to stockpile, so the company can operate on a steady basis. To keep up with demand requires a total crew of eight who work in shifts, so processing runs on a 24-7 schedule, plus five contract haulers are "kept busy daily hauling from the storage at Waseca." Of the eight employees, only one lives in the township, on his farm northwest of the village.

On their biggest single day, they loaded out twenty-one cars of finished product, which required that a switch engine had to come from Waseca with seven empty cars three times and place them on the siding and take seven loaded cars to the larger siding at Waseca so the shipment could go on one train. A good business has been established, with Fleet Farm being one of their largest customers. It is reputed that New York City alone requires fourteen carloads of bird seed daily to keep the bird watchers happy. In 2010, the only business in the village had a through put of 55 million pounds of just sunflower seed.

Marion Davidson became the village postmaster when Anna Drache retired in 1968, and after all the businesses had closed, except the Peterson Grain Company, the federal government closed the post office on October 2, 1992. On March 12, 2000, Wade Homuth, who owned the bank building that had housed the post office since 1968, razed it for safety reasons.

Because of the large volume of business that the village with only one firm gave the railroad, it named a locomotive for it. Taken at Brookings, SD, in 2004 by Cody Grivno, Editor *Model Railroader Magazine.*

Meriden lost its final business landmark when what had served as the Kath store since 1964 was destroyed by fire June on 14, 2003. Eddie Kath confirmed that there was evidence of a robbery that resulted in arson. This building dated to 1879. After the fire, the house, which had served as post office and the Paul and Anna Drache home, now became the oldest structure in the village. The house, located across from the Peterson Grain Office, was built in 1868, with additions in 1930 and 1939. A small depot, which was erected in 1866 and was expanded twice, was taken down in about 1960. Sampson's Dairy Products provides rural delivery of milk, bread, pastries, and other food, so those who live in the village have access to that service. RFD out of Waseca and Owatonna provides mail service. Due to environmental regulations, the village was compelled to install a sewer system. The village is no longer a farm service center. In addition to the Peterson Grain plant, the only remaining vestige of the once active business district is the town hall.

It would have been too costly to convert the school house to make it handicap accessible, so the building was demolished and the two-acre school lot became the site for a new town hall and garage. The trees were planted in 1923.

The only reminder of any business other than farming that once operated in the village is the Yule Truck Line at Medford. When Yule purchased the Drache Truck Line in 1968, he acquired ten customers in Owatonna who had been with Drache. Yule's business grew to a very sizable operation as those customers expanded. In 2009, one of those businesses was no longer operating but the remaining nine were still with him. Members of the Eddie Kath family have several ventures in Owatonna, and members of the Albert E. Born Well Drilling enterprise in Waseca have grown that to one of the largest of its kind in southern Minnesota.

Where's Meriden?

As I commented in the Preface, it was not until 1992 that I decided to write a micro-macrostory, i.e., what was taking place in the township and how was it affected by what took place beyond its border. Even before I did any writing, I had observed how my Dad's business evolved from livestock trucking to hauling freight for businesses in Owatonna. It was clear to him that was where the greatest opportunity lay. After we returned to Concordia College to teach, we continued to farm and most of our social life was with people in the country. We always listened to the farm markets. The only organization I joined other than church was the Northwest Farm Managers Association, which had been started by managers of the bonanza farms under the guidance of one of the pioneer agricultural economists in the nation. They were a very upbeat group and so positive about agriculture–a sharp contrast to the traditional "poor farmer" harangue that we heard on the radio when we were trying to learn what the market was doing. It was much like the steady "how bad things are" we heard from some farmers.

In the late 1960s, when I was researching for my second book about homesteaders in the fertile Red River Valley, I realized that 160 acres were not sufficient to provide an adequate family life style. That is what compelled me to continue my research. I decided I wanted to find out why some farmers were always so positive while others were always complaining. One of the books I researched had a statement from the president of a national farm organization speaking at their annual meeting. He told his audience, "Do not forget when you get home be sure to remind others that farmers are second class citizens." I continued my search. I wrote to a United States Senator and asked if he could give me a direction. He replied with a printout of all of the top income farmers based on their earnings from the government programs. That list became my guide to who I would interview for my book on larger-than-average farmers. None of them ever asked why I had selected them, but most of them were of the entrepreneurial bent and realized how to capitalize the program proceeds into expanding their operation. At the same time the politicians campaigned on the slogan that they were all out to save the small farms and small towns of rural America, even though they knew that cheap food was a key to the well being of any nation and we were leading the world in that respect. That research led to two books: one on how technology helped change agriculture and the other about the psychology of successful farmers.

Wheeler McMillen, whom I have quoted previously, wrote that farmers were the primary makers of America's wealth. Agriculture has been one of the leaders in maintaining a positive balance of trade. Earl Heady asked, "Do we want to help feed the needy of the world or do we want to preserve our former way of farming?" But when I learned that in 1928 the members of the Gavel Club of Owatonna were aware that the excess labor released by a changing agriculture would be there for their future needs, I fully realized what my theme would be. Today, if you want to buy a candy bar or a bottle of pop in Meriden, you will have to go to the vending machine in the grain company office or have it delivered by the weekly food delivery truck. My little home town has served its purpose and is no longer needed. I have heard many former farmers say the good old days are today.

Statistics show that it is not unreasonable to ask the question. When the New England frontier was first settled, a village was established in every township as a means of protection from the Indians. As the nation expanded, farming progressed beyond the subsistence level, and a village was established about every eight miles to cope with the existing mode of travel.

By the mid-1930s, travel conditions improved and the rate of mechanization accelerated. At that time we had 6.8 million farms, and today we have 1.9 million, of which the top 20 percent produce more that 85 percent of the total output. As average farm size increased 10-fold, farm population fell from 30 million to 2 million. But every year the nation had an abundant supply of healthy, low-cost food. This came about because in 1862 Abraham Lincoln encouraged an equally foresighted Congress to enact legislation that led to an entirely new direction for agriculture. Those acts, combined with improved technology and entrepreneurial farmers, brought changes that did not require a village every eight miles. Most of those villages have outlived their usefulness, and we are justified in asking "Where's Meriden?"

The Daily People's Press, 12, 16, 24 September 1977, 2 July 1983, 5 July 1995, 12 March 2000, 16 June 2003; Paul Drache Day Book; Timothy Arlt, telephone interview, 27 January 2012; *Chronological Landmarks* 85-89; Dan Peterson, personal interview and telephone conversations, 28 July 2006; Drache, E. M., *Young Prairie Pioneer*, 121, 122 ; Drache, *Plow Shares* 97; Kenneth Dinse, personal interview; *Minnesota Agricultural Statistics 1983*, 3; Andy Cummings, DM&E Rides Off Into the Sunset, *Trains* Dec. 2007: 10–11; McMillen, *Too Many Farmers*; Earl Heady, "Small Farm Systems," 67–91.

REFLECTIONS ON LIVING THE DREAM

The story concentrated on the settling of a township and the rise and decline of a village was created to serve those who opened the sod and strove to make their contribution to society. In previous chapters, Veblen was cited that in the small farm villages there was always someone there to create a "boom" in order to make a profit and then leave. He was critical of those individuals but at the same time he wrote that the majority of the small town merchants did not do well financially. In Meriden there were three persons who might have been considered boomers. They were F. W. Goodsell, H. C. Palas, and E. L. Scoville. In 1903 the daughter-in-law of Goodsell and her daughter moved to a new home in Waseca. At that time they owned some lots in the village and 176 acres in section eighteen. Palas had sold much of his land by 1914, but he continued to live in Meriden. The Scoville family never had more than two quarters of land and I remember them as living modestly. It was previously mentioned that in 1868 Goodsell built the largest home in the village at a cost of $600. That house was purchased by my parents in 1923 and served as the Meriden post office from 1928 to 1968. It still stands.

No business person who started in the village after 1900 accumulated any sizeable fortune. Those who lived out their business life there worked long hours and seldom took much time off, but they were able to live comfortably in retirement. The greatest turnover was the store west of the county road and in the pool hall/barbershop. Veblen was correct that most small town businesses never had enough customers so the turnover rate was high, but apparently that was considered normal.

The following is the story of a couple who farmed for forty-four years—my aunt and her husband. I worked for them for two summers in the 1930s so I knew them well.

They probably were more typical of the farmers in the township than the farmers previously described as leaders in farming practices. They never had any children, and I might have filled a void in their lives at least for two years. To my knowledge the only other hired help they ever had was for grain harvest and corn husking. In chapter five it was mentioned that when they were married and started farming, Max Drache furnished them with livestock, machinery, hay, and seed that totaled $2,570. This was a loan and was paid back.

Leo Lebahn was born in Owatonna in 1900, and after he finished eighth grade he worked on farms in Meriden Township. Aunt Mollie was born in 1901 and worked at home until 1918 when she was employed to work at the William Scholljegerdes farm, her mother's brother. She earned fifty cents a day when she started, which was gradually increased to seventy-five cents a day, room and board included. By 1922 the Scholljegerdes children were able to help on the farm so she was no longer needed.

From 1922 until she was married in September 1926, she worked at home where there was "plenty to do," for in addition to a large flock of chickens, there was a large flock of geese, a herd of hogs, and about twenty cows plus the small dairy stock and horses. Leo continued working on nearby farms and owned a Model T pickup. When they were married, Mollie received a set of household furniture that included a dining room set, two bedroom sets, a kitchen stove, and a kitchen set, for which her parents had paid $100. When she died in 1979, I remember that, except for the kitchen stove, she was still using that furniture. The kitchen table has the typical enamel top, not quite as large as two card tables with a very small drawer and small, square, white wooden legs. The only other pieces of furniture that they had were rocking chairs and an enclosed phonograph record stand with a phonograph on the top shelf. I think I played it twice on rainy days in the two summers I worked there.

After her parents died, the pump organ on which my father had learned to play was moved to her house but never played. Our son has it now. In 1926, they rented 160 acres in section twenty-one from Max Drache for $600 cash. In addition to the machinery mentioned in chapter five they purchased a used walking plow for $15, a used wooden Boss harrow for $5, and before harvest in 1927, a new McCormick eight-foot binder for $250. That binder harvested every crop they raised. Later, they purchased a used dump rake and a new McCormick hay loader for $100. They purchased two more horses in 1930, and in 1936, like almost every farmer in the township, lost a horse to sleeping sickness. They never owned any other horses, but in 1939 they purchased an International "H" and a tractor plow and corn planter for $950.

They milked twelve to fifteen cows, which required five to seven minutes per cow at each milking. They separated the milk and used a "milk cart" to take the skim milk to the hog barn. They only had two cream separators, both used, during their forty-four years of farming. I will never forget that I always had to keep my mouth closed tightly when I separated because the flies were so thick. The same was true when I dumped the skim milk into the swill barrel in the hog barn. From 1927 through 1961 their smallest cream check was $12 and the highest $125.

In 1962, they built a new barn for $2,237 completely equipped, including a milk-cooler. After that they sold whole milk for which they received from $100 to $225 per month. In addition to the cows, they had four sows every year that farrowed twice yearly, except in 1932. The pigs were fed ground oats mixed with skim milk and finished on ear corn. They sold twenty-five to thirty hogs per year. Each year they butchered a "big fat sow." The biggest was 630 pounds which produced twenty-three gallons of lard. In addition to the pork they fattened some bull calves and each year kept one for butchering so they could have a quarter of beef and sold the rest.

A chicken flock of 300 rounded out the farming operation. Some years they had as many as fifty clucks to hatch the new flock because the hens were butchered, dressed, and sold every year. In 1960, they purchased an incubator for hatching. The chickens provided four eggs every day of the year for eating in addition to fried chicken dinners. They never owned a refrigerator, but they used the milk cooler in the barn to keep some food there in the warm months.

In 1945, they paid $650 to install electric lights with Interstate Power. "It was the biggest change that took place in our farm life when we put in lights. We used it for pumping water, the washing machine, and for the corn and grain elevator. In 1967 after our hands got so sore that we could not milk any more we got a milking machine, but we never owned a refrigerator. Our first light bill was $5.05." In 1968, they had their largest annual gross income, $9,900. In 1970, they still used a threshing machine and had to hire people to help thresh because the other farmers all used combines. In 1973, their threshing machine was sold for $25 to the KOA campground south of Owatonna.

In the fall of 1970, they sold the cows for $3,463.62. The granary was full of grain so they kept the chicken flock, but they quit farming and rented out the land. My uncle died in 1974 after which my aunt had an auction that yielded $3,000. They had purchased the tractor and the Packard car with money from an inheritance, and Mollie stated that they had about $8,000 in the bank when Leo died.

During their life they had a Model T pickup for their first car and in 1929 they purchased a four-door Model A Ford for $520 and in 1939 a Packard for $1,000. They had three other pickups—an International in 1935, a 1942 Dodge, and a used Ford in 1968, for which they paid $250.

They were content with life, but like the merchants in the small town, they worked long hours and I never recall that they ever took a vacation. They did their weekly shopping and attended church. After Leo died,

Mollie had her Social Security income and money from renting out the farm, and she fed the grain that was left in the granary to the chickens. She also sold the eggs except for the two she ate each day. She told me that she tried to live on just the Social Security income "because that is what it was for" and saved the rest.

Prior to Leo's death I had spoken to them about buying the farm and told them that they could live on it as long as they wanted, and they agreed with that. After he died she was plagued with realtors who told her, "Mrs. Lebahn you have got to sell the farm and move to a more comfortable place in town." Adeline Drache Evans, her sister and my youngest aunt, called me and said, "You better come down here because someone is pestering her to buy the farm nearly every day." Ada and I traveled to Owatonna and the first thing Mollie said was that she did not want to leave the farm. She had no attorney and would not go into town to see one and asked me to call the one that my parents had used. He understood and willingly came to the farm and with Adeline, who was her power of attorney, met around the little white enamel table in the kitchen. At that point she had $14,000 in the checking account and the farm.

The farm was located adjacent to the city limits, which made it choice property. It was worth more than the total profits they had made in their life time. She told me how much she wanted for the farm, and I agreed with her offer. The county treasurer told me that it was the highest price that had been paid for any farm land in Steele County up to that date. Mollie was happy, for she lived in her home until the last ten days of her life when she had to be taken to the hospital.

I have thought about their lifetime of farming often since 1974 when I purchased that farm and compared it to what I had gained from my experience in farming and all those I have interviewed who were leading agriculture into a new era. Leo was not a master farmer by any means. He never joined DHIA to improve his herd, nor did he take part in any meetings to learn about changes that were taking place in agriculture. He certainly never contacted the county extension agent for help. He owned a tractor after 1939, but he had a team of horses almost to his final days of farming. He might have used hybrid seed corn but I am not aware that he ever adopted any of the innovations, in his lifetime of farming, including something as basic as commercial fertilizer. I never heard them complain about how bad things were regardless of what the prices were, and never once did I hear them feel sorry for themselves. I am certain that Mollie had it better than many of those who had businesses in the small farming communities across rural America. She was living in her own home and had land to support herself, but what did the small town merchants have when they could not find a buyer for their business?

Since the career of the Lebahns, even though the nation has lost several million farmers, production has continued to rise. Other developed nations, which have adopted agricultural technology and consolidated their farms, have had the same experience. Therefore, the need for little farm service centers, like Meriden, will diminish.

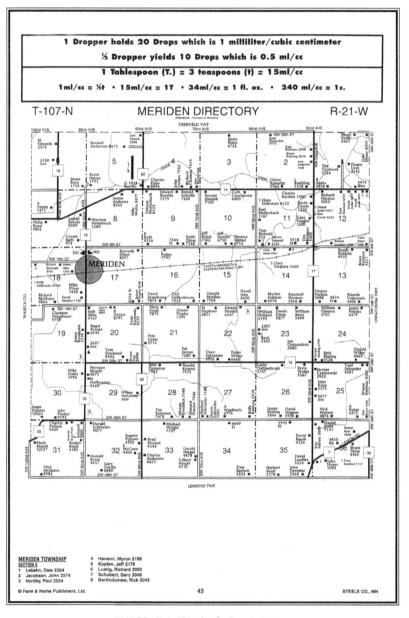

2009 PLAT A GIS - Steele County image.

APPENDIX

Meriden Lists

CHICAGO NORTHWESTERN RAILROAD EMPLOYEES:

AGENTS:

1885 - A. G. Bossard
1891 - 1895 - Freeman O. Biennis
1892 - James Brydon
1896 - L. W. Schnee
1903 - 1904 - Charles N. Nickolson
1903 - John G. Burns
1903 - 1906 - Edward W. Hefferman
1905 - O. F. Thomas (new depot)
1907 - M. Hosfield
1907 - Guy Cotton, housing shortage
1908 - V. V. Vine
1908 - J. J. Schekaw
1909 - A. C. Booner
1910 - G. E. Cox
1911 - E. C. Miekke
1911 - Gordon W. Ware
1912 - Al Beyers & Floyd Wilcox
1913 - J. L. Westrom
1915 - 1918 - Paul Sellnow
1918 - A. J. Hammel
1918 - 1919 - G. G. Gucholz
1919 - 1926 - Alonso Sawyer
1926 - 1928 - Gaskel Warren
1928 - Frank Gresch
1928 - Neil C. McDougall
1930 - Clifford Lundholm
1931 - 1935 - Elmer Johnson
1935 - 1939 - Dave Piche,
1939 - 1963 Lloyd Grandprey caretaker; then Waseca took over.

FOREMEN:

1882 - 1904 - Pat Maher
subs: 1891 - Pat Miner; 1901 - Joe Herzog
1901 - Wm. Smith or Albert Smith
1906 - 1911 - Frank Thieves
1912 - 1915 - Matt Thieves
1916 - Casper Rahling
1923 - ____ Watson
1924 - 1934 - Floyd Walters
1931 - 1932 - R. C. Lee
1932/35 - 1953 - John Evans
1953 - Raymond Evans, in 1956 the
Waseca, section moved to Waseca.
Evans retired 1974

LUMBERYARD OWNERS AND MANAGERS:

Hitchcock
Otto Nietardy
    1895 - C. A. Litchfield
    1896 - Otto & Paul Nietardy
Winona Lumber Co.
    1897 - Hayes Lucas Lumber Co.
Managers
    1903 - August Schmidt
    1903 - Herman Martin

1908 - 1916 - Ernest Schuldt
1916 - 1918 - Louis Domy, he was drafted in WWI.
1918 - 1925 - Emil Buboltz
1925 Ernest Schuldt was moved to Owatonna.
1925 - 1963 - Lloyd Grandprey, when Hayes Lucas sold out.
1959 - Harold Schroeder established Schroeder Cashway Store in the former pool hall/barbershop which sold hardware items and fertilizers.

COMMUNICATION:

1866	- CNWR had Western Union and American Express.
1889	- The Wicklow Store and Hayes Lucas had telephones for public use.
1896	- An Owatonna telephone in the Post Office lobby.
1911	- Waseca telephone in Luhman Store for public use.
1913 - 1961	- The Meriden Rural Telephone Company initially had two circuits to Owatonna.
1920	- The Waseca Bell Company had a line to the township.

POSTMASTERS:

Postmaster	*Appointment Dates*
Marcus C. Flower	December 22, 1856
Joseph Guenther	December 18,1860
Frederick J. Stevens	February 12, 1861
Archibald M. Dickey	January 12, 1869
Frank M. Goodsell	June 22, 1870
Wilson E. Widger	December 11,1874
Franklin W. Goodsell	May 27, 1875
Adolph W. Peters	September 1898
Samuel E. Grandprey	March 26, 1904
Henry J. Luhman	November 26, 1917
Mrs. Anna L. Drache	December 3,1928
Mrs. Marion M. Davidson	April 30, 1968
Closed	October 2, 1992

BLACKSMITHS:

Prior to 1870 - A Mr. McCall left Meriden to move west.
1870 - 1875 - Peter Pump then went farming.
1883 - E. A. Stebbins.
Wm. Engle sold his shop in 1893.
1891 - N.G. Pratt built a shop north of tracks east of the county road.
1894 - Sam Grandprey took over Pratt's shop.
1895 - E. E. Hocking built on Herman Stendel lot, 2 stories MWA hall, burned in 1899.
1896 - John Franz partner of Grandprey until 1905.
1897 - Grandprey Hall/shop was built on Mill Street.
1898 - James Maloney, worked for Grandprey.
1899 - Fred Fleischer had a shop in section two near Golden Rule creamery.
1901 - Julius Meitzner worked with Grandprey.
1902 - Charles Markmann 1917 worked with Grandprey.

1904 - Herman Gertje with Grandprey and Markmann.

1909 - Bill Haas joined Grandprey and Markmann

1915 c.a. - Fred Radke worked either independently or with Markmann but left sometime in the 1920s for Cresbard, South Dakota, and returned to Owatonna in the 1930s after a couple years to open a shop in Owatonna. This writer remembers his shop on North Cedar Street.

1915 - Grandprey, sold the Hall and shop lot to the bank group for $600 and moved the shop north of the tracks near the creamery and added twenty feet to the west end of the building.

1917- Grandprey left Meriden and Haas and Markmann took over.

1919 - Julius Meitzner took over the business and in 1928 went farming.

1928 - 1981- Dick Rietforts, he was 65 in 1961 and reduced work as the business declined.

1983 - The building was moved to Farm America west of Waseca.

ELEVATOR AND MILLS:

The original building was a flat storage structure that was erected soon after 1866 by the railroad or a line elevator company.

1873 - E. C. Strong & Co., J. R. Fox, agt.

1879 - Meriden Mill Co., J. Z. Barncard

1879 - F. W. Goodsell to Meriden Mill Co. (along track/Grandpreys)

1883 - A tornado caused severe damage and it is not certain that the mill operated again.

1883 - E. S. Wheelock, Hitchcock, & A. Simpson

1885 - Meriden Mill Co. moved to Lake Kampeska, So.Dakota

1891 - E. L. Scoville, flat grain storage operated by a one-horse sweep to load grain to the train.

1903 - Hastings Mill, John Franz, mgr.

1904 - Sheffield King Co.

1906 - S. E. Grandprey, manager for Pride Elevator Co. or for Hastings Mill.

1911 - L. G. Campbell Milling Co., makers of Malt-O-Meal, S. E. Grandprey, manager.

1916 - Meriden Farmers Elevator organized by farmers, J. C. Wilker, pres., Herman Schuldt, sec. 1916 - Leo Haas manager, 1917 - Juluis Bartz manager until March 1921.

1921 - The Meriden Farmers Elev. & Mercantile Co. was formed in April. Paul Drache was manager until he resigned in 1925, and in 1926 the company went into bankruptcy.

1926 - Hayes Lucas purchased the firm in July.

1926 - Thomas Clarke leased it in August until May 1927.

1927 - Art Willert became the owner/operator.

1932 - Herman Thordson leased until February 1933.

1933 - 1934 - Herman Stendel ground feed on an irregular basis under Willert's direction.

1934 - 1937 - Paul Drache rented the warehouse and sold concentrates and other feeds that he purchased to provide backhaul loads for his trucks.

1937 - Henry Olson and his brother, who had other elevators, completely remodeled, updated, and expanded the facility. The Meriden Grain Co. was a very successful business.

1955 - 1969 - Owatonna Farmers Elevator expanded the facility with Duane Miller as manager.

1965 - Harold Schroeder started in grain the buying business.

1969 - Schroeder purchased the Owatonna Farmers Elevator business. January 7, 1970 - On January 7th the main structure was consumed by a fire. Schroeder continued to buy grain until he was caught in the middle of Hedge-to-Arrive dealings and was forced to quit.

1971 - 1985 - Midland Coop, International Minerals Corp., Darrell Johnson & George Jones were involved for various periods.

1985 - Dan Peterson bought out Victoria Grain, Land-O-Lakes, which had the west end fertilizer shed, and the big grain bin from Allen Terpstra and established Peterson Grain Company.

GARAGES:

1917 - L. C. Palas north of the tracks east of the county road; Harvey Palas operated it until he retired about 1975.

1923 - Fred Fette, south side of Mill Street east of the county road operated by sons Herb and Ernest; after 1935 Ernest had a radio and electrical sales and repair shop in the barn on the Fred Fette home site.

1942 - Drache Truck Line purchased the Fette garage and operated it primarily for company business but sold gasoline and oil and did minor repairs for the public.

1948 - Hiram Drache purchased the garage and leased it to Corwin Kanne and then to Ewald and Herman Zacharias.

1957 - Herman Zacharias purchased the business and ran it until his death.

1965 - Kath Brothers

1967 - Mike and Louise Stenzel purchased the building and leased it to Jim Stockwell for two years. Then Stenzel operated it when he was not busy with his full time job until 2011 when it became a private storage only.

1967 - The creamery was converted to a garage and Jim Raetz used it for his construction business.

John Pipers rented it to make specialty castings.

George Hardy rented it for selling used cars and to do general repairing. He made an apartment on the second floor.

HOTELS:

1879 - 1884 - Central Hotel, above the Herman Rosenau store and saloon, located south of the tracks west of the county road, operated by George and Barbara Hubbard.

1896 - S. E. Grandprey rented the hotel to Charles Depue.

1898 - Mrs. Parcher, daughter of Grandprey, operated as Meriden Hotel for boarders and roomers.

1898 c.a. - Henry Luhman purchased the building for his store and the hotel and meals were operated by Mrs. William Walthers.

1906 - Meriden Hotel - William Walthers leased it for a brief period.

Then the second floor was occupied by the Luhman family and apartment dwellers.

GENERAL MERCHANDISE STORES:

West End:

1886 - Cannot identify

1892 - J. R. Patrick & L. C. Wolter

1897 - L. C. Wolter & Albert Speckeen

1898 - H. J. Luhman (telephone station) In 1917 he became the postmaster and built a 20 x 70 foot store building; had the office in the store.

1905 - Isker & Wolter, H. H. Wicklow mgr.

1936 - 1942 - Homer Donaker, Self-Serve Market Edwin & Ellen Spiekerman managers.

1942 - 1946 - Herb & Leona Pankow

1946 Robert Wesley Bros. & Deml

East End:

186?- E. E. Goodsell & A. M. Dickey

1870 - E. E. Goodsell & Waumet

1885 - J. R. Petrik & E.W. Clark

1891- Petrik & Albert Speckeen

1906 - H. H. Wicklow to Aug. 22, 1945

1945 - Leslie McGray

1946 - Diemer & Lil Davis

1973 - Jeff Kath purchased the building and converted it to apartments.

1948 c.a. - Clifford & Alvina Hofius

1949 - 1954 - Bill Gunderson

1954 - 1964 - Leo Thompson

1964 - 1976 - Eddie & Kay, after which they sold only confectionary items but used the building until it burned June 14, 2003, arson.

There were two other businesses that could not be identified - 1891 Fred Mudeking place of business and in 1915 Frank O'Rourke general merchandise.

DEVELOPERS, CARPENTERS, & PAINTERS:

1889 - D. C. Ross

1890 - 1913 - John Korupp and sons Fred and Henry were the leading carpenters in the area during this period. They built the three local creameries and several others in the area and were the leading home builders.

1896 - Chris Tage

1898 - Samuel E. Grandprey built three houses north of the tracks west of the county road.

1900 - Henry Brase

1905 - Adam Walther

1914 - For several years, Harry Grandprey, Lloyd Grandprey, Alvin Buelow, Leslie Leader, and Paul Schipple worked with various carpenters.

1915 - 1940s - Rudolph and Gustav Schendel did carpentry and painting.

1916 - L. C. Palas built several houses in Meriden and some on farms.

1920 - 1923 - Ewald Fette built several on the east end and also north of the tracks.

1929 - 1940s - Herman Stendel built chiefly farm buildings.

1925 - 1940s - Bill Ahlers, a township native but lived in Owatonna, was a major builder.

194? - Frank Raetz built many structures in the township for about twenty years.

196? - James Raetz used the creamery for his construction business

BARBER SHOP & POOL HALL:

1911 - Harry Grandprey barber and harness repair

*** G. E. Grosse barber

1916 - 1924 - Jack Osborne barber. The school building was moved east to provide space for the Fette garage and became the pool hall/barbershop and later it was enlarged to become a family dwelling.

1924 - 1930 - Harvey & Grace Moore barber, pool hall, tobacco, confectionery, and lunches.

1930 - 1934 - Bill and Rose Arndt barber, pool hall, tobacco, confectionery, and lunches.

1934 - 1942 - Harvey & Grace Moore barber, pool hall, tobacco, confectionery, and lunches.

1942 - 1947 - Vergil & Viola Hires Harvey & Grace Moore barber, pool hall, tobacco, confectionery, and lunches.

1952 - 1958 - Bill Christenson

1952 - 1960 - Harvey Moore came out to barber one night a week at Voss's Three Vs.

From 1952 to the 1970s Don Emard and Gayhart Johnson came out to Renchins one night weekly to barber but dates are not certain.

EATING PLACES, CONFECTION, BEER, ET AL:

1887 - Theodore Fedder paid $100 for a saloon license.

1914 - Mr. Bartels had a wholesale liquor license.

1926 - Ernest or Harold Brase opened the Midway at the corner north of the village initially as a service station along with soft drinks and candy. It was destroyed by fire.

1930 - Herman Brase rebuilt the station.

1931 - 1943 - Al Bell's Midway, as a service station, candy, soft drinks, beer, and reputedly "something to add zip to the beer."

1943 - 1951 - Al Lutgens as Al's Midway, as a service station, candy, soft drinks, beer, and reputedly "something to add zip to the beer."

1943 c.a. - Ernest Fette moved his electrical business to the open lot on Mill Street located between the post office and the large Fred Fette house, and that evolved into hamburger/beer business. It was sold to Merel and Isabell Kleist who operated it as a short-order, beer, and pool hall.

1947 - 1951 - Anna Drache financed M & M Shop, which was a lean to on the east side of the Drache garage. It specialized in short orders and was operated by Helen Schoonover & Rosina Schultz. Then it was rented to Mildred Zacharias who ran it as Zack's Cafe until November 1951.

1951 - 196? - Gay Krenke at the junction corner service station with beer, candy, and short orders.

1952 - 1960 - Walter & Gertrude Voss purchased the Kleist building and operated as the Three Vs "Voss's Vitamin Vineyard," appropriately named by R. J. Stuart, a leading customer. They specialized in hamburgers and beer, short orders, pool tables, and a juke box. One night a week a barber was on duty.

1960 - 1985 - Albert and Marie Renchin purchased the Three Vs. They sold the property to Peterson Grain, which purchased it to gain access to the well and razed the building.

196? - Lowell Brooks had the corner station, which was burned on a Christmas Eve but they continued to live in the adjacent dwelling.

2011 - There was no gasoline, bread, or candy available in the village other than the vending machine in the Peterson Grain Company office.

MISCELLANEOUS BUSINESSES:

1880 - Charles Gee hardware and tin shop

1880 - L. C. Wheelock well driller

1882 - Abernathy Brothers well driller

1887 - Theodore Fedder saloon

1889 - Mr. Shipman took orders for boot and shoes.

1894 - William Dietz shoemaker

1896 - A. Warren well driller

1899 - William Stoltz owned a custom threshing rig and had the small farm at the east end.

1902 - L. C. Palas custom threshing and shredding rigs

1903 - E. Bonnel feather renovator which were used in mattresses and pillows

1905 - T. E. Rogers Watkins dealer

1905 - McConnon & Co. dealer not identified

1906 - 1916 - Flavia Peters stenographer

1915 - Art Hammann lived in Owatonna but for years drilled many wells

1916 - Richard Wells Watkins dealer

1928 - 1952 - Albert E. Born: 1928-1932 Waltkins dealer; 1928-1952 well driller (later and sons)

1955 - Myron Davidson Raleigh Products dealer

CROWN CREAMERY: (Located in section twenty-five)

Year	Pounds of Butter	Income	Butter Maker
1894	7.593	$1,805	John Stensval
1895			S. G. Kinney
1900	151,719	$31,288	fire destroyed the building
1906	95,633	$22,661	
1912	102,646	$30,377	
1913	114,118	$35,571	
1915	121,865	$35,555	
1919	135,723	$68,036	
1921	162,855	$65,700	
1923	198,031	$85,718	
1924	197,808	$82,196	Paul Gertje 1924 c.a.
1925	209,562	$91,682	
1926	198,000	$85,458	
1928	190,185	$87,997	
1929	196,739	$85,568	
1930	193,694	$67,464	
1931	185,799	$49,450	
1933	207,186	$48,481	
1935	186,170	$54,375	
1936	207,669	$67,854	
1938	217,308	$58,422	
1939	218,043	$55,003	
1940	246,260	$71,564	
1942	215,834	$84,742	Closed 1947

GOLDEN RULE CREAMERY: (Located in section two)

Year	Pounds of Butter	Income	Butter Maker
Started April 1892			W. I. Noy
1894	13,752	$3,053	
1898	***	***	
1905	67,990	$14,719	
1906	57,899	$13,090	George Deeg
1909	62,388	$15,386	Stensval
1910	71,192	$21,704	
1913	80,475	$25.345	George Deeg
1916	86,809	***	
1921	102,145	$40,617	
1925	128,591	$57,253	
1927	129,639	$60,107	
1928	116,205	$54,253	
1930	119,474	$42,194	
1931	120,966	$32,175	
1935	128,098	$37,585	
1938	159,733	$44,059	Closed 1949

MERIDEN CREAMERY: (Located in section eighteen)

May 1, 1891, creamer started by E. J. White

Year	Pounds of Butter	Income	Butter Maker
1895	A cooperative association		Henry J. Rosenau
1900	177,460	$33,597	
1905	193,337	$43,743	
1908	162,289	$42,804	
1911	194,307	$46,452	
1912	194,588	$61,192	
1913	195,134	$61,604	
1915	212,193	$66,511	
1916	207,475	$71,072	1917 - John Grosser
1918	237,178	$115,200	
1919	245,567	--	
1920	254,619	$163,079	
1921	292,255	$128,505	
1922	345,786	$142,925	
1923	328,038	$187,401	
1924	402,191	$162,838	The third largest creamery in Steele County.
1925	396,221	$182,997	Purchased a buttermilk drier and received four cents a pound for dried buttermilk, O. P. Jensen.
1926	378,121	$174,182	
1927	386,304	$189,845	
1928	342.548	$165,482	
1929	341,525	$158,332	
1930	346,447	$110,808	December Jensen resigned, Loren Luhman hired.
1931	369,154	$93,319	
1932	401,271	$81,377	
1934	427,464	$109,746	
1935	394,330	$117,673	
1936	428,575	$143,279	
1938	447,585	$123,832	
1940	489,713	$143,830	
1941	320,000		Started buying whole milk and enlarged the creamery for more tanks.
1942	Loren Luhman to Owatonna and Arthur H. Uecker employed.		
1944	344,090		
1948	--	$265,759	
1952	512,617		
1956	807,958		Purchased 21,000,000 pounds of milk
1958	881,919	$717,583	
1959	952,642		
1965	1,657,424		Had 170 members
1966	The creamery shut down and all milk was hauled to Owatonna.		

The Price of Butter Fat for Various Years—Cents per Pound

1901 - $0.22	1917 - $0.55	1921 - $0.35	1933 - $0.27
1903 - $0.19	1917 - $0.57	1921 - $0.41	1934 - $0.33
1906 - $0.21	1918 - $0.72	1923 - $0.47	1938 - $0.32
1908 - $0.30	1918 - $0.81	1926 - $0.62	1941 - $0.41
1910 - $0.39	1919 - $0.84	1927 - $0.50	1943 - $0.46
1911 - $0.36	1919 - $0.60	1930 - $0.44	1945 - $0.54
1912 - $0.42	1919 - $0.88	1932 - $0.25	1947 - $0.82
1917 - $0.52	1920 - $0.79	1932 - $0.19	

YEAR		BIRTHS	DEATHS	YEAR		BIRTHS	DEATHS
1872	TWP	--	--	1913	TWP	21	9
	CTY	215	73		CTY	388	128
1874		8	1	1916		25	5
		299	104			--	--
1878		33	10	1917		19	9
		356	107			--	--
1880		39	6	1920		21	5
		398	125			380	170
1882		33	6	1924		15	6
		335	159			351	176
1884		24	14	1925		13	8
		282	130			385	191
1886		20	11	1926		19	3
		--	--			330	166
1888		30	5	1929		13	5
		339	101			327	172
1894		28	5	1932		18	3
		101	43			348	173
1905		17	4	1933		9	4
		336	122			301	163
1907		20	2	1934		5	6
		360	157			315	165
1908		13	8	1935		9	3
		345	155			337	179
1910		19	6	1940		2	2
		352	169			403	203
1912		14	4	1945		--	---
		377	169			436	171

POPULATION: FARMS:

Year	Twp	Cty	Twp	Cty
1857	-	493		
1866	233	2,363		
1870	739	8,288		831
1875	898	10,759		
1880	809	12,488	220	977
1885	833	12,752		
1890	834	13,232		
1900	880	16,524		1,801
1910	805	16,146		1,824
1920	764	18,061	184	1,860
1930	808	18,475	140	1,925
1940	755	19,749		1,970
1950	746	21,155		1,875
1960	832	25,029		1,715
1970	791	26,931		1,404
1980	719	30,328		1,067
1990	693	30,729		819
2000	631	33,680		774
2010	621	36,576	51	934

An increase in small farms occurred because the gross scale figure was reduced. A farm is any unit that has $1,000.00 in gross scales.

FARM COMMODITY PRICES:

Date	Wheat	Corn	Hogs	Beef	Butter	Eggs
1871	0.86	0.60			0.12	0.12
1874	0.91	0.50			0.25	0.20
1880	0.94	0.23	3.25	3.70	0.15	0.18
1882	1.26	0.65	6.10		0.10	0.10
1886	0.66	0.33	2.75	3.00	0.10	0.19
1893	0.54	0.30	6.00	2.00	0.25	0.22
1898	0.80	0.25	3.10	2.75	0.12	0.17
1905	0.78	0.28	4.25	3.75	0.25	0.19
1910	1.02	0.45	8.45	5.00	0.31	0.26
1915	1.19	0.52	6.25	6.50	0.34	0.30
1921	1.50	0.27	5.75	3.25	0.40	0.36
1924	1.23	0.70	9.75	7.50	0.35	0.27
1928	1.12	0.40	7.00	7.00	0.44	0.25
1932	0.49	0.30	2.20	4.75	0.21	0.09
1935	0.89	0.76	8.67	8.75	0.28	0.20
1940	0.65	0.51	4.70	8.40	0.28	0.13

The dollar was devalued in 1932, which accounted for the rise in the year that followed. The prices came from the local papers and if there was a range in price quotations I used the average of the high and the low.

LAND PRICES:

Date	Size of Farm	Price/Acre	Section
1855	The established price by law was $1.25 per acre		
1875	160	$16.25	20
1877	160	$12.50	14
1883	160	$24.10	11
1884	160	$41.50	24
1890	160	$18.44	33
1892	80	$32.50	34
1894	80	$62.50	22
1896	160	$31.25	19
1898	190	$35.26	7
1898	240	$60	20 Henry Palas to Wm. & Lewis
1899	80	$50	10 Abbe a well-developed farmstead.
1903	120	$62.50	2
1907	120	$62.50	16
1909	120	$100	16
1911	240	$75	36
1913	120	$100	27
1914	200	$220	10
1914	160	$106.25	21
1919	240	$225	31 In Clinton Falls twp adjacent
1920	160	$275	35 In Otisco twp adjacent
1925	160	$125	27
1927	200	$125	34
1929	200	$125	30
1941	315	$539	13 The Gilkey Farm

PRICE OF VILLAGE LOTS & HOMES:

YEAR	PRICE	LOCATION
1885	$300	In the business district
1886	$350	
1887	$3,620	All the property north of the tracks on the west of the county road except the first three lots owned by Joseph & Samuel Grandprey and 150 feet owned by the creamery, 100 feet owned by Henry Rosenau, and a lot owned by Korrup. This property was divided into eight lots including the farmstead just north of the creamery.
1894	$210	East end 8 rods x 8 rods
1896	$700	Frederick Walter to L. W. Schnee; this was a large lot, and the latter had a large white house, barn, and small buildings and became the Fred Fette home.

YEAR	PRICE	LOCATION
1897	$800	Eva Fedder (Mrs. Theodore) to A. W. Peters, a lot and building south of the tracks from the county road east to the rear of the Peters store which became the backyard of the Wicklow store, bank, and pool hall/barbershop.
1898	$1,000	Otto Nietardy to Winona Lumber Co. sub lot 2 in lot 6.
1901	$950	Hayes Lucas to Bertha Buelow sub lot 2 in lot 6; later this became the house where the Hayes Lucas manager lived.
1901	$1,050	Johannes Nietardy to Rudolph Dietz a lot on the hill east of the county road and south of the tracks south of the first school house.
1901	$50	Goodsell to Sarah Parcher the lot for the third house north of the tracks west of the county road and south of where the town hall/blacksmith was later built.
1901	$100	Goodsell to Peter Pump a half acre for a garden for the third house north of the road that went west out of the village. It is still a garden lot in 2012.
1902	$1,000	E. L. Scoville to Gustav Schendel west half of sub lot 6.
1907	$1,700	F. Karsten to Rudolph Schendel on the north end of the village and west of the county road a residence a barn and four acres.
1907	$2,000	H. F. Martin to L. H. Schultz a lot at the east end of the village south of the tracks.
1914	$2,550	Ernest Schuldt to Herman Stendel south of the tracks east of Wicklow store large lot with a fine house and a barn.
1915	$650	The lot where the Grandprey hall and blacksmith shop were located between the Wicklow store and the pool hall/barber shop for a bank.
1915	$2,950	P. Gertje to Fred Fette the land west of where the Drache home/post office was later located until 1968 and east to the Gus Schendel lot.
1916	$2,300	Ernest Schuldt to Louis Domy with a barn and a substantial house located second house east of the Wicklow store, it still stands (2012).
1918	$6,000	Cordes to Theodore Pump ten acres on the east end with a substantial house and two small buildings.

1918	$1,700	Bertha Buelow to Hayes Lucas Company east end which was remodeled for the company manager, from 1925 it was home of the Lloyd Grandprey family until it burned in 1977.
1918	$2,800	Joseph Grandprey to Lena Drache this included a barn and a substantial house just north of the tracks west of the county road.
1918	$3,300	Louis Domy to Wilheminia Beier with barn, chicken coop, and wood shed and a house that still stands second one east of the former Wicklow store.
1923	$4,500	? Gertje to Paul Drache with a large house a barn and two acres.
1925	$1,000	to Mueller Brothers with a residence and two sheds.
1934	$450	Reuben Eggers to Max Drache a lot, five rods by eight rods, and a residence. Drache added a garage/wood shed and later Raymond Evans lived there.
1934	$5,500	Ewald Dinse to Paul Drache east end farm with a fine house, a barn, five other buildings, and fifteen acres of land.
1935	$3,000	Henry Papke to Dick Rietforts house built in 1922.
1953	$2,000	Raymond Evans to Elmer Kruesel (see 1934 above) who razed the house and built a new one.
1970	$16,000	Paul Drache to Frank Dinse the east end farm he purchased in 1934 (see above).
1973	$10,000	Lillian Davis to Jeff Kath, the store building which was converted to apartments.
1977	$7,000	Grandprey estate to Eddie & Kay Kath, the lot that held the house that Hayes Lucas had purchased for its managers and Lloyd & Eva Grandprey had live there from 1925 until 1977. Kath then built a new house on the premise.
1988	$27,000	Paul Drache to John Staley the house Drache purchased in 1923 (see above) to which they enclosed north porch, added a garage for two trucks, cemented out the barn, moved in two woods sheds and converted them to garages for cars. In 1939 they removed the original kitchen and built a larger kitchen/dining area plus two rooms and a bath on the second floor. About 1970 they put in a new wall in the basement at a cost of $7,000 and when natural gas became available they installed a gas furnace.

DEERFIELD MUTUAL:

Founded April 4, 1881, by 400 policy holders i.e. members who were motivated by the Great Chicago Fire of October 1871 which resulted in huge losses to many insurance companies so the locals decided to establish their own farmers mutual.

	Policy holders	*Insurance in force*
1890	654	$910,207
1891	1,076	$1,484,626
1892	1,388	-0- , paid $5,100 in losses
1895	1,593	$2,263,577
1897	1,928	$2,772,045; losses $3,551, a farm must be at least 40 ac. to be a policyholder.
1900	2,129	$3,391,320; losses $5,114
1903	2,463	$4,533,914; largest single loss $945, a granary; J. H. Wilker elected president.
1906	2,430	$4,623,860; losses $6,276, John Eliasen V. P.
1907	2,469	$5,167,450; losses $8,411, covers all of Steele County, four townships in Dodge & Waseca.
1908	2,498	$5,507,249; losses $16,980: $8,150 lightning, $8,830 fire
1910	2,496	$5,554,460; losses $3,713
1911	2,551	$6,479,129; losses $10,600; moved to insure cars just like horses; a necessity, lost 89 to 62
1912	2,586	$6,978,268; losses $15,721; Gust Fette $2,805, a large barn, some cattle, all hay & grain
1913	2,635	$7,249,414; losses $9,504; $1,846; largest loss a barn, all hay, grain, and straw pile
1918	-0-	-0- added livestock & cars to fire and lightning on crops & bldgs.
1919	-0-	$13,826.368; 72 losses, $17,043; largest loss $4,293 for a barn, silo, & hen house; motion to increase salaries defeated.
1923	3,265	$14,206,233; losses $28,464, most to fire
1924	3,278	15,345,837; cannot insure in villages
1925	4,082	19,930,163; car and truck insurance only for travel in Minnesota
1926	3,711	16,324,903; 500 attended meeting in court house
1927	-0-	21,309,484; made agreements with five area towns to pay for fire runs
1928	3,190	21,119,709; after a speech by county agent it was agreed to insure purebred cows up to $300
1929	3,905	22,925,266; the $1.00 service charge on delinquent payments was successful
1930	3,953	23,870,256; total losses $43,360; continued working with five area towns
1931	3,254	Losses $38,700
1932	3,268	19,430,689
1933		J. H. Wilker, president since 1903, died
1935	3,644	Losses $32,486
1936	3,728	19,726,200; 164 losses $19,504; voted to have 5 directors
1938	4,435	23,865,406; losses $25,415; discussed owning a fire truck; Owatonna Rural Fire Dept. will charge $30 for home member calls
1940	3,748	20,316,919; losses $27,701; rural truck made 12 calls
1946	3,590	Losses $22,878; compulsory inspection for electrical wiring
1947	3,436	24,947,470; losses $20,608
1949	3,529	Losses $37,499

1951	3,241	$32,258,650; the reduced number of policies was possible personal property and buildings were consolidated into one policy
1964		Name changed to Owatonna Mutual Fire Insurance Co.
1967	2,620	$44,180,300; the smaller number of policy holders reflected the decline in farms in the area

THE MERIDEN LIVESTOCK SHIPPING ASSOCIATION:

Year	Members	Head	Cars	Gross Income
1910	190	3,218	48	$59,521
1911	205	6,168	98	$90,003
1912		7,368	102	$111,656
1913		7,804	108	$128,742
1914		7,456	123	$138,194
1915		8,663	145	$143,479
1916	2,150,000 lbs.		128	$146,022
1917	2 Horses	8,620	137	$234,981
1918	67 Members	8,134	123	-0-
1919	10 Horses	9,213	140	$337,440 Fred Fette resigned and Theo. Pump managed.
1920		7,847	115	
1921		7,045	115	$132,585
1922				Members could not vote if they had sold live-stock outside of the association.
1923		8,176	117	$124,383
1924	9,039 Hogs	10,669	148	$149,503
1925		7,477	114	$167,700 First time no sheep; telephone $81.25.
1926	4,667 Hogs	5,732	84	$140,447
1927	One Sheep & 4,386 Hogs			$105,420
1928	2,163 Hogs	2,679	49	----
1929	1,443 Hogs	1,723	42	$40,669

EARLY CATTLE BUYERS:
Probably the first cattle buyer was Dr. Fordyce Brown, who was a Civil War veteran and based on his obituary "lived in the township several years" and apparently dealt in livestock for resale. He died in 1896 in Michigan.

1890 E. L. Scoville
1891 Scoville & Schuldt dissolved; Scoville quit
1895 Anton Schuldt, Will Warren
1895 J. H. Schuldt (Ernie's dad)
1895 L. G. Wolter & A.W. Peters
1896 Hastings, Scoville, & Wolters
1906 Wm. Walther

FIRST STATE BANK OF MERIDEN: Organized July 1915
Wilker, J. C Wilker, J. H. C. Schuldt, Fred Domy, Ernest Schuldt, H. J. Luhman, H. W. Schuldt, H. J. Rosenau, S. E. Grandprey, C. F. Markmann, H. H. Wicklow, L. H. Abbe, Fred Mundt, H. W. Papke, Gus Fette, William B. Voss, Peter Moe, Joseph Karsten, Max Drache, C. H. Wilker, S. W. Kinyon, C. J. Kinyon, John Ebeling, J. W. Ebeling, A. G. Schmidt, F. C. Buscho, Henry Boege, H. J. Karsten, J. L. Westrom, E. J. Buscho, E. G. Papke, H. C. Abbe, L. Olhoeft, Michael Krause, Paul Krenke, Theodore Pump.

Year	Assets	Dividends
1915	$35,598	Capital stock $10,000
1917	$134.109	
1919	----	Capital stock $25,000; 65 stockholders.
1920	$300,630	
1921	$304,750	
1926	$630,240	
1929	$604,283	$24,000 was paid in dividends in the ten years from 1920 to 1929 and dividends were paid every year until 1959.
1932	$457,500	
1933	$212,541	
1934	$214,641	
1935	$253,437	
1937	$319,151	
1938	-0-	$1,500 W. P. Jones held over 51 percent of the stock.
1939	$262,294	$1,500
1940-44	-0-	$14,500
1945	$38,552	$7,500
1946-48	-0-	$19,500
1949	$611,520	$5,000
1951	$725,955	
1952	$939,788	$2,500
1953	$890,595	$2,500
1954	$950,709	$2,500
1955	$819,987	$2,500 Only 18 stockholders.
1958	-0-	$649
1959	$576,161	---- Lost $402.44, the only loss, and the last year in Meriden.
1960	$596,411	Reorganized as Oakdale State Bank in Owatonna.
1968	$4,101,699	The final year that anyone from Meriden was involved.

TEACHERS:

Dist. 10/2089	Dist. 14/2093	Dist. 21/2098
No dates Given	*Established in 1856*	*All prior to 1879*
Jeff Green	1911-13 Luther White	Emma Davis
Clementine Murphy	1917 Mabel Berg	Miss Leighton
Victoria Green	1918 Frieda Reiman Bredlow	Jenny Harris
Dora Wood	1929-31 Caroline Helegson	Mary Harsha
Ophelia Aldea	1936 Grace Woods	1895 Alice M. La Salle
Susan Martin	1937 Marion Benike	1899 Mrs. Hardy
	Caroline Betlach Woker	1903 Eva Ross
Mr. Sanborn	1940 Laurietta Miller Mc-	1904 Hulda Peters
S. B. Williams	Nearney	1910 Beatrice Burns
1870s James Harris	1948-50 Mrs. Albert Grant	1911 Mal Christianson
1870s Carrie Fredenburg	1950-51 Ada Marie Drache	1913-16 Mary Welch
1978 Nellie Schlieman	1952-53 Pelagia Kosmoski	1924 Margaret Ringhofer
1879 Ada Chick	1954-58 Margaret A. Shea	1925 Bertha Larson
1891 Bertha Warren	1959-60 Olive Kelley	1926 Miss Bruzek
1892 Ella Carter	1961-62 Evelyn Ruzek Ille	1929 Ellen Eggboe
1894 Bertha Warren	Closed	1931-35 Melvin Jenke
1895 Clara Ebner		1939 Katherine Thykeson
1913 Tessie Tulley		1941 Lillian Davis
1923 Mildred Yust		1957 Lorranine Anderson
1925-28 Martha Ringhofer		1960 Carmen Meyers
1929 Roselyn Thamert		1966-68 Leona Stenzel
1930 Wilma Hostad		Closed
1931-33 Edith Jerouseh		
1939 Emma Baker		
1941 Mildred Hallberg		
1957 Iriewa Newgard		
1966-67 Wanda Taschner		
Closed		

Dist. 36/2111	Dist. 39/2113	Dist. 43/2116
1923 Dora Houdek	Prior 1879 Emma Sheldon	1896 Clara Ebner
1924 Mrs. Caroline Helgeson	1894 Rachel Peters	1899 Miss McRostie
1925-28 Jessie Lawson	Luther Richardson	Miss Weber
1929 Mrs. Caroline Helegson	1912 Martha Aurauff (?)	1905 Gladys Locke
1930-31 Marion Duncan	1913 Ruth Barker	1907 Margaret Loone
1932 Georgia Kubick	1923 Marie Jirele	1910 Ida Zimmerman
1933 Arlyn Ellerman	1925-26 Ruth Jones	1911 Ethel Clark
1934 Elsa Lachmiller	1927 Emma Kriesel	1912 Mary M. Barnes
1935 Pearl Remund	1928 Elizabeth Ringhofer	1913-14 Mary Welch
1936 Ella Casperson	1929 Emma Kriesel	1917 Eleanor Withers
1941 Leona Ringhofer	1930 Blanche Nass	1919 Amelia Ryshaug
1950 Ingrid Hanson	1931 Martha Ringhofer	1923 Miss Erickson
1957 Clara Carlson	1932-37 Gladys Olson	1924 Ruth Barker
1960 Marie Ballstadt	1938 Lorna Blume	1925-26 Grace Johnson
1966-68 Marie Ballstadt	1939 Edith Bartsch	1927-29 Margaret Peterson
	1941 Mildred Hankerson	1930 Louise Markson
	1957 Clara Barton	1932-35 Lorna Blume
	1960 Glenda Barton	1936 Verna Briese
	1966 Lillian Wardelman	1938 Elizabeth Crandall
	1967-68 Doris Fogal	1941 Adeline Bettschen
	Closed	1943 Elizabeth Gasner
		1957 Verna Briese
		1960 Verna Briese
		1966 Marcell Hanson
		1967-68 Dorothy Powell
		Closed

Dist. 52/2121	Dist. 92/62	Dist. 78/2147
1857 Dianty Leroy	1911 Mary Welch	1892 Fannie Grandprey
1879 Ellen Williams	1912 Ella Drum	1893 Miss Backus
1891 Eila Andrew	1912 Mabel Christenson	1915 Medora Grandprey
1898 Daisy Warner	1913 Beatrice Blazek	1920 Eva Ochs
1912 Miss Sanders	1916 Gerda Preus	1928 Blanche Nass
1914-15 Eva Ochs	1917 Miss Woods	1933 Elvera Schmidt
1923 Miss Braden	1925 Ruth Barker	1938 Gladys Johnson
1925 Mildred Nass	1927 Bernice Remund	1939 Lillian Dau
1926-27 Irene Nieb	1928 Mrs. Wells	1957 Hildegard McGowan
1931-33 Blanche Nass	1930 Hildegarde Glock	
1935 Elevera Oeltjenbruns	1931 Irene Lien	
1936 Lorean Nelson 1937	1932 Magdalen Kvasnicka	
1936 Isabel Jensen	1935-36 Violet Beese	
Dist. 16	1936 Elizabeth Crandall	
1890 Fanny Grandprey	1938 Bernard Kramer	
1892 Miss Backers	1939-41 Pauline Pirkl	
1894 Frank Warenett	1942 Leona Pankow	
1901 Elizabeth Patton	1957 Shirley O'Conner	
1902 Hulda Peters		
1914 Medora Grandprey		

TOWNSHIP CENTURY FARMS: As of 2009

OWNER	*YEAR*	*SECTIONS*	*ORIGINAL OWNER*
Mike Dinse	1866	17 & 18	Henry Abbe
Jim Andrews*	1856	5 & 8	Thomas Andrews
Weldon Beese	1884	4	Peter Eliason
Kenneth & Helen Brase	1856	30 & 31	Christopher Wilker
Richard V. Ebeling	1894	24	Michael Ebeling
Mary Ebeling Pichner	1869	32	Michael Ebeling
Gerald &Berniece Schroht	1878	13	Ludwig Reiter
Gerald & Berniece Schroht	1894	23	F. W. Schuldt
Emil Steinberg Family	1892	34	Allen Radel et ux**
Chester Waumett	1855	3	John Waumett
Vernal Wilker	1878	21	Christopher Wilker

*Also 150 years. In 2004 all but fifteen acres of this farm were auctioned off to Radel and now is owned by the Joe Stransky family.

A PARTIAL LISTING OF TOWNSHIP OFFICERS:

When the Lloyd Grandprey home was destroyed by fire in 1977 virtually all the township records were destroyed, so this listing will at least make some of those names available.

Year

1857   The final proceedings by the county commissioners were authorized and the following were listed as early active officers: F. J. Stevens, A. F. Tracy, Samuel Reemsnyder, J. O. Waumett, T. P. Jackson, E. L. Scoville, W. F. Drum, W. T. Drown, Joseph Grandprey, Henry Leroy, E. L. Crosby, Robert Stevenson, L. G. Green

1877   Supervisors John Martin, Peter Beck, W. F. (or H) Hobbins, assessor F. B. Davis, treasurer F. W. Goodsell, clerk James A. Harris

1878   Supervisors E. L. Scoville, Byron Gillett, W. F. Hobbins, clerk James Harris, assessor F. B. Davis, treasurer F. W. Goodsell, justice of peace (hereafter justice) F. W. Goodsell, constable Fred Steffen

1880   Supervisors E. L. Scoville, J. D. Backus, Byron Gillett, clerk James Harris, treasurer F. W. Goodsell, assessor W. F. Hobbins, justice O. Abernathy, Wm. Thompson, constables Major A. O. Rowley, Charles Geo

1882   Supervisors E. L. Scoville, John Wilker, Peter Beck, clerk A. O. Rowley, treasurer F. W. Goodsell, justice O. Abernathy, constable Albert Sanders

1883   Justice of peace Abernathy resigned and E. S. Wheelock appointed, H. Schuldt resigned as constable and Wm. Engle appointed

1885   Supervisors John Wilker, Henry Drinken, H. N. Lutgens, clerk J. C. Burke, treasurer F. W. Goodsell, assessor W. F. Hobbins, constable H. Stendel

1887   Supervisors Fred Walther, J. D. Backus, H. Rosenau, clerk Wm. F. Hobbins treasurer F. W. Goodsell, assessor R. G. Rosenau, justices Joseph Grandprey and F. W. Goodsell, constable Henry Stendel

1889   Supervisors J. D. Backus, H. Rosenau, A. C. Warren, clerk W. F. Hobbins, treasurer F. W. Goodsell, assessor R. G. Rosenau, justice Joseph Grandprey, constables H. Stendel, A. Speckeen

1890   Supervisors John Wilker, Henry Brase, Gottfried Bosshard, clerk W. F. Hobbins, treasurer F. W. Goodsell, assessor James Harris, justices F. W. Goodsell, P. McRostie

1891   Supervisors John Wilker, H. Lutgens, Herman Rosenau, clerk W. F. Hobbins, treasurer F. W. Goodsell, assessor R. G. Rosenau

1892   Supervisors J. W. Martin, H. Lutgens, Henry Drinken, clerk W. F. Hobbins, treasurer E. E. Goodsell, assessor Frank Beckman, justices F. W. Goodsell, Samuel Grandprey

1893   Supervisors H. W. Lutgens, Henry Drinken, R. Andrews, clerk W. F. Hobbins, treasurer Louis G. Wolter, assessor Frank Beckman, justice James Harris, constables Samuel Grandprey, W. Abbe

1896   Supervisors Frank Beckman, J. W. Andrews, H. Rosenau, clerk Lyman Warren, treasurer Louis G. Wolter, assessor Adolph Peters, justices Sam Grandprey, F. W. Goodsell, constable Samuel Grandprey Jr.

1899   Supervisors J. W. Andrews, Wm. Walker, John Aberling, clerk Adolph Peters, treasurer Henry Drinken, assessor W. F. Stoltz, justice Peter Pump, constable Charles Rotke

1901   Supervisors John Ebeling, E. Horton, Peter Pump, clerk H. F. Korupp, treasurer Henry Drinken, assessor H. L. Walther, justice H. Brockmiller

1903   Supervisors Fred Fette, Peter Pump, Herman Rosenau, clerk H. F. Korupp, treasurer Sam Grandprey, assessor Walter Joriman, justice W. F. Hobbins, constables Wm. Hundt, Ed. Buelow

1904   Supervisors Fred Fette, E. J. Horton, F. W. Schuldt, clerk H. F. Korupp, treasurer Theodore Pump, assessor Walter Joriman, justice John Ebeling, constables E. Lutgens, Herman Stendel

1905   Supervisors Fred Fette, E. J. Horton, F. W. Schuldt, clerk H. F. Korupp, assessor Walter Joriman, justice Wm. Woker, constable Adam Walthers, general fund $800

1906   Supervisor A. S. Smith, clerk H. F. Korupp, treasurer Peter Pump, assessor Walter Joriman, justice H. Lutgens, constable Mike Ebeling

1907   Supervisor Mike Ebeling, clerk H. F. Korupp, treasurer Peter Pump, assessor Walter Joriman, justice Carl Ebeling, Wm. Woker, constable Henry Rosenau, general fund $2,000

1910   Supervisor J. C. Wilker, clerk H. F. Korupp, treasurer Sam Grandprey, justice Wm. Brase, constable A. H. Wilker, general fund $2,000

1911   Supervisor Ben Kuchenbecker, clerk H. F. Korupp, treasurer Sam Grandprey, assessor Walter Joriman, justice Herman Brase, constable Henry Rosenau, general fund $2,400

1913   Supervisor Theodore Pump, clerk Ernest Schuldt, treasurer Sam Grandprey, assessor Wm. Woker, justice Herman Brase, constable Henry Rosenau, general fund $2,200

1914   Supervisor J. C. Wilker, clerk Ernest Schuldt, treasurer Henry Drinken, justice John Westrom, constable Alvin Buelow, general purposes $500, poor fund $50

1915   Supervisor Herman W. Schuldt, clerk John Westrom, treasurer Henry Drinken, assessor Otto Ebeling, justice Fred Fette, constable Henry Rosenau, general fund $3,700

1916   Supervisor Theodore Pump, clerk Herman Brase, treasurer Henry Drinken, justice Albert Wilker, constable Alvin Buelow

1918   Supervisor Wm. Brase, clerk Herman Brase, treasurer Henry Drinken, justice Julius Bartz, constable Louis Beese

1921   Supervisor Wm. Brase, clerk Herman Abbe, treasurer Henry Drinken, justice Nicholas Phillips, constables Frank Miller, Art Brase, general fund $5,000

1922   Supervisor Albert Wilker, clerk Theodore Pump, treasurer Henry Drinken, general fund $5,000 of which $3,500 was for road and bridge

1924   Supervisor H. C. Abbe, clerk Theodore Pump, treasurer Henry Drinken, justice Paul Drache, constable Ludwig Schultz, total tax levy $35,050 of which $5,000 was general budget, $4,486 for roads, and $9,507.52 for schools

1925   Supervisor Herman Brase, clerk Theodore Pump, treasurer Henry Drinken, assessor Wm. Woker, justice Paul Drache, constable Rueben Luhman, general budget $6,000 of which $4,500 was for roads

1927   Supervisor Fred Altenburg, clerk Theodore Pump, treasurer Henry Drinken, assessor Wm. Woker, justice Paul Drache, constable L. C. Palas, general fund $8,000 of which $6,000 was for roads

1928   Supervisor George Reul, treasurer Henry Drinken, justice Art Brase, constables Paul Krenke, Herman Schuldt, general fund $10,000

1929   Supervisor Wm. Brase, treasurer Paul Drache, assessor Wm. Woker, justice Art Willert, constable Paul Krenke, general fund $7,000

1960   Supervisor Elmer Krause, clerk Art Willert, justice Lloyd Grandprey, constable Melvin Abbe, zoning program approved, general fund $5,000

1971   Supervisors Norbert Abbe, Bill Gleason, treasurer Arnold Abbe, general fund $20,000

1972   Supervisors Robert Armstrong, Joe Houdek, clerk Lloyd Grandprey, justice Lowell Brooks, general fund $20,000

BIBLIOGRAPHY

Public Documents

Ahearn, Mary, et al. *Farming-Dependent Counties and the Financial Well Being of Farm Operator Households*. USDA-ERS, Agricultural Information Bulletin No. 544. Washington: GPO, August 1988.

Bender, Lloyd, et al. *Diverse Social and Economic Structure of Nonmetropolitan America*. USDA-ERS, Rural Development Research Report No. 49.Washington: GPO, September 1985.

Minnesota State Land Office. *State Land Office Approved Lists: R.R. Land Grants 1860–1936*: SAM 45, roll 57, book 5, and R.R. *Land Grants deeded* February 12, 1867. Minnesota Historical Society: State Archives.

North Dakota State University, Department of Agricultural Economics. *Industrialization of Heartland Agriculture: Challenges,Opportunities, Consequences, Alternatives*. Ag ricultural Economics Miscellaneous Report No. 176, December 1995. Fargo: NDSU.

Steele County, Minnesota. *Book of Original Entries, Plat of 1879*. Owatonna.

USDA. *Report of the Commissioner of Agriculture for the Year 1862*. Washington: GPO, 1863.

Articles, Bulletins, Cassette, Letters, Minutes, Periodicals, Newspapers

Abbe, Esther. Letter. 6 June 2005.

---. *Genealogy of the Wilker, Scholljegerdes, Drache Families.*

Andrews, James M. Letters. 25 May 2005, 14 November 2008.

Baker, O. E. "Farm Youth, Lacking City Opportunities, Face Difficult Adjustment," *Yearbook of Agriculture*, 1934. Ed. Milton Eisenhower. Washington: GPO, 1934. 207–209.

Cummings, Andy. "DM&E Rides Off Into the Sunset." *Trains* December 2007: 10–11.

District Reunion Committee. "History of the Meriden School 48/2118." June 1989.

Drabenstott, Mark, and Tim R. Smith. *The Changing Economy of the Rural Heart Land*. Federal Reserve Bank of Kansas City, April 1996.

Drache, Anna. Recollections of living in Meriden from 1923 to 1968.

Drache, Hiram M. *The Life and Times of Paul and Anna Drache: As it Relates to the History of the Community of Meriden, Minnesota.* Typed manuscript, 1988.

Drache, Max. Farm record booklet, 1913–1934.

Drache, Paul. Day books on the operation of the Drache Truck Line and personal memories, 1 January 1937–6 October 1989.

Durand, E. Dana. *Cooperative Livestock Shipping Associations in Minnesota.* University of Minnesota Agricultural Experiment Station Bulletin No. 156. St. Paul: 1916.

Elazar, Daniel J. "Federal-State Relations in Minnesota: A Study of Railway Construction and Development." Diss. U. of Chicago, 1957.

Ellsworth, Clayton S. "Theodore Roosevelt's Country Life Commission." *Agricultural History* XXXIV, No.4, October 1960: 155–172.

Fette, Mildred. Life in Meriden. Cassette tape, 10 March 1996.

First State Bank of Meriden. Minutes, 1915–1959.

Froelich, Martin. Letter. 11 May 2005.

Grandprey, Hattie O. Hersey. A Memoir, c.a. 1900.

Heady, Earl O. "Small Farm Systems." *Proceedings of the Lucas Memorial Symposium on Systems Concepts in Agriculture.* North Carolina State University, March 1980. 67–91.

Hugdal, R. M, Acting Inspector in Charge. Letter to postmaster, Meriden, Minnesota. 28 December 1928.

Hudson, Dorothy Wicklow. Letter. 17 November 1999.

Johnson, Charles (aka Kenny). Letters. 7 July 1992, 9 February 2005.

Kennon, Peggy. Cassette tape with Dick Rietforts for Farm of America Pioneer Village. 8 January 1986.

Kriesel, Emma. *A Wilker Family History*, c.a. 1950.

Manke, Lorna Blume. *Teaching and What Else: Rural Schools of Yesteryear, Steele County Minnesota.* Owatonna: The Rural Teachers' Committee, 1979.

McCoughlin, D. M., postmaster Waseca, letter to the Honorable M. I. Ryan, Inspector in Charge, St. Paul. 10 November 1942, in possession of the author.

Meriden Creamery Association. Minutes, 17 October 1938–11 May 1960.

Meriden Farmers Elevator & Mercantile Company. Audit report, 31 May 1924; final audit, 31 July 1926.

Meriden Rural Telephone Company. Minutes, 5 January 1913–1961, in files of Steele County Historical Society.

Minnesota Agricultural Statistics, 1983.

Minnesota Railroad & Warehouse Commission. On proceedings of closing the CNWR depot in Meriden.

Mosher, Jonnie. *A Brief History of the Pump Family.* October 2003.

Rietforts, Gertrude. Letter. 10 May 1989.

Schoenfeld, Mary Jo. Letter and Century Farms data, 16 October 2009.

Schroeder, Harold. Typed manuscript of his recollections. 22 October 2004, 12 February 2007.

Steele County Landowners of Record, 1879. Complied for the Steele County Historical Society, 1988.

Steele County Photo News, 30 June 1938–1975.

Successful Farming. *Greatest Tractors of All Time.* Vol. 1. Successful Farming, 2011.

The Owatonna Journal/Owatonna Journal Chronicle, 1868–January 1938.

Truesdell, A. J. *Directory of Steele County, Minnesota*, 1892–1893.

Uecker, Arthur. The operation of the Meriden Creamery 1938–1960. Cassette tape, 20 July 1999.

USDA-ERS. *Chronological Landmarks in American Agriculture.* Agriculture Information Bulletin No. 425. Washington: GPO, 1975.

Waseca Campaign. August–November 1872.

Waseca County Landowners of Record, 1879. Compiled for the Waseca County Historical Society, 1982.

Waseca Herald, 15 December 1961.

Waseca News, 10 October 1867–25 December 1872.

Williamson, Harold. "Mass Production for Mass Consumption." *Technology in Western*

Civilization, Vol. I. Oxford: 1967.

Zelinski, Marian Schoonover. Letter. 1 October 1997.

BOOKS:

Andreas, A. *Illustrated Historical Atlas of the State of Minnesota*. Chicago: A. T. Andreas, 1874.

Berg, Walter G. *Buildings and Structures of American Railroads, A Reference Book for Railroad Managers, Superintendents, Master Mechanics, Engineers, Architects, and Students*. New York: John Wiley & Sons, 1893.

Casey, Robert J. & W. A. Douglas. *Pioneer Railroad: The Story of the Chicago and the North Western System*. New York: McGraw Hill Book Co., 1848.

Chicago Farm Equipment Institute. *Land of Plenty*. Chicago: Farm Equipment Institute, 1959.

Child, James A. *History of Steele County, Minnesota: An Album of History and Biography*. Chicago: Union Publishing Company, 1887.

Danhof, Clarence H. *Change in Agriculture*: The Northern United States, 1820–1870. Cambridge: Harvard University Press, 1969.

Drache, Hiram M. *Beyond the Furrow*. Fargo: North Dakota Institute of Regional Studies, 1976.

---. *The Challenge of the Prairie*. Fargo: North Dakota Institute of Regional Studies, 1971.

---. *The Day of the Bonanza*. Fargo: North Dakota Institute of Regional Studies, 1964.

---. *History of U.S. Agriculture and Its Relevance to Today*. Danville: Interstate Publishers, 1996.

---. *Koochiching*. Danville: Interstate Publishers, 1983.

---. *Plowshares to Printouts. Danville*: Interstate Publishers, 1985.

---. *Tomorrow's Harvest*. Danville: Interstate Publishers, 1978.

Draffan, George. Taking Back the Land: *A History of Land Grant Reform*. Seattle: Public Information Network.

McMillen, Wheeler. Too Many Farmers: *The Story of What is Here and Ahead in Agriculture*. New York: William Morrow & Company, 1929.

Shapsmeier, Edward, and Frederick H. Shapsmeier. *Encyclopedia of American*

Agricultural History. Westport: Greenwood Press, 1975.

Severson, Harold C. *The Night They Turned on the Lights: The Story of the Electric Power Revolution in the North Star State.* Privately published, 1962.

"The Agricultural Adjustment Act, May 12, 1933." *Documents of American History.* Ed. Henry Steele Commager. New York: Appleton-Century-Croft, 1958. 422–426.

Veblen, Thorstein. "The Country Town." *The Portable Veblen.* Ed. Max Lerner. New York: Viking Press, 1948. 407–430.

Woolworth, Alan R. *The Genesis & Construction of the Winona & St. Peter Railroad, 1858–1875.* Marshall: The Society for the Study of Local & Regional History at the History Center, 2000.

INTERVIEWS:
Andrews, James W. Meriden. Letters and several calls, 2006–2010.
Arlt, Timothy. Mankato. Telephone, 27 January 2012.
Born, Donald E. Waseca. 8 September 2008.
Buecksler, Paul F. Owatonna. 8 July 1989.
Dinse, Kenneth. Meriden. Several conversations and calls, 2006–2012.
Dinse, Sandra. Meriden. Several conversations and calls, 2006–2012.
Drache, Anna L. Meriden. 11 September 1976.
Drache, Paul A. Meriden. 11 September 1976.
Eggers, Alice. Owatonna. 9 September 2002.
Eggers, Clarence. Owatonna. 9 September 2002.
Eggers, Judy. Owatonna. Telephone, 20 December 2011.
Fette, Mildred. Owatonna. 14 June 1996.
Grandprey Jr., Lloyd. Faribault. Several calls, 2006–2010.
Grandprey Sr., Lloyd. Meriden. 11 September 1976.
Grivno, Cody. Slinger, WI. Telephone, 20 September 2011.
Hudson, Dorothy. Stratford, Iowa. Various dates.
Jensen, Bernice. Owatonna. Telephone, 20 December 2011.
Kath, Edward. Owatonna. 8 September 8 2008 and several conversations.
Kopischke, Rosina Wilker Schultz. Waseca. Telephone, 20 May 2009.
Krause, Loren. Owatonna. Telephone, 28 January 2012.
Krenke, Margaret. Owatonna. Telephone, 6 July 2009.
Lebahn, Dale. Meriden. Telephone, 20 June 2010.
Lebahn, Mollie Drache. Owatonna. 5 January 1972.
Mosher, Jonnie. Owatonna. 11 October 2003.
Murray, Gladys. Frederick, Maryland. Several calls, 2006–2010.
Palas, Virena Stendel. Waseca. 14 June 1996.
Peterson, Dan. Meriden. 28 July 2006, and several conversations.
Rietforts, Dick. Meriden. 21 October 1985.
Ruttan, Vernon W. St. Paul. 20 April 1970, 24 April 1978.
Scholljegerdes, Robert, Waseca. Letter, 29 January 2002.
Scholljegerdes, William E. Meriden. 7 July 1979.

Schultz, Robert. Owatonna. Telephone, 16 May 2011.
Stenzel, Mike. Meriden. Telephone, 21 February 2012.
Voss, Gertrude. Meriden. 16 July 2002.
Willert, Charles. Nekoosa, Wisconsin. Telephone, 2 July 2009.
Willert, Hilda. Meriden. 11 September 1976.
Yule, Curtis. Medford. 20 May 2009.

INDEX

ORIGINAL ENTRIES 1856–1858

MERIDEN T.107N. – R.21W

6	5	4	3	2	1
Mathias Malget James Brown Peter Beck	Israel Curtis John Bradley	Robert Andrews Thomas Bradley William Webster	John R. Allen Samuel Erskine John O. Wuamett Harriett Latro	Luther Huntsley David A. Foote A. C. Hams	Orville M Ford Luther Huntsley William Bryant, Jr. Albert McKinney

7	8	9	10	11	12
David Howard James Dirkson	Richard Larrington John Oleson Mathias Malget Elijah Ash	Charles H. Mashan William Hallow	Lorenzo Green John D. Tutthill	Gale Williams John Schwieso	George Show Gale Williams O. N. Crane

18	17	16	15	14	13
Edwin Hunt Ole Nelson Charles Wentworth Emory North	Jesse F. Keel John H. Foot Samuel Adams Emory North	Sarah Waramett	Aaron Matson John Glover Solomon Wetzell Hiram C. Thompson	William Thompson Henry Bradley William O. Goltry Aaron Bascom	Hiram M. Sheetz Harriet A. Sheetz Thomas Treat Merritt H. Middaugh

19	20	21	22	23	24
William R. Taylor John Dunken Charles Wentworth	Henry Peterson Oscar Crandall William Brown William Swain	Edward Higgis Alva E. Daggett George W. Crane	John A. Armstrong Joseph Craig John Grear Charles Kitchen	Michael Colger John Middaugh, Jr. Ai Middaugh George E. Leary	Helen M. Lambert James W. Kerns Samuel G. Epla Hardin Andrus

30	29	28	27	26	25
Henry Sanneman Howard Hatch Christopher H. Wilker Thomas Sullivan Henry Sanneman Heinrich Sanneman Ernest Sanneman	Henry Abbe Anton Schuldt	Heinrich Abbe George House David House	Patrick Collins Frank Seaver David House	Stillman Washburn Enos Kelsey Lysander House Salmon B. Washburn	Eliher W. Rogers Russell Town Zerah A. Town

31	32	33	34	35	36
Henry S. Clement William F. Drum F. J. Stevens C. H. Wilker Carl Schuldt	Anton Schuldt John Passow Christian Blohm	John Duckening Richard Whitlock Jackson F. House Daniel House	John Duckening	Michael Barney Francis Delong	

2009 PLAT OF MERIDEN

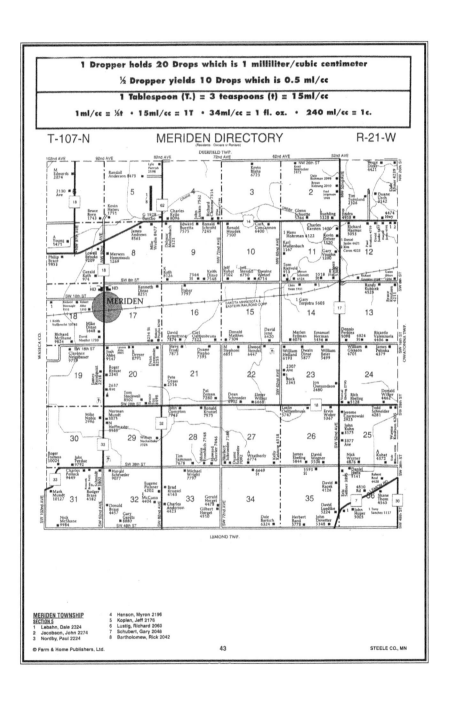

1 Dropper holds 20 Drops which is 1 milliliter/cubic centimeter
½ Dropper yields 10 Drops which is 0.5 ml/cc
1 Tablespoon (T.) = 3 teaspoons (t) = 15ml/cc
1ml/cc = ⅓t • 15ml/cc = 1T • 34ml/cc = 1 fl. oz. • 240 ml/cc = 1c.

T-107-N   **MERIDEN DIRECTORY**   **R-21-W**

MERIDEN TOWNSHIP
SECTION 5
1 Lebahn, Dale 2324
2 Jacobson, John 2274
3 Nordby, Paul 2224

4 Hanson, Myron 2196
5 Koplen, Jeff 2176
6 Lustig, Richard 2060
7 Schubert, Gary 2048
8 Bartholomew, Rick 2042

© Farm & Home Publishers, Ltd.   STEELE CO., MN

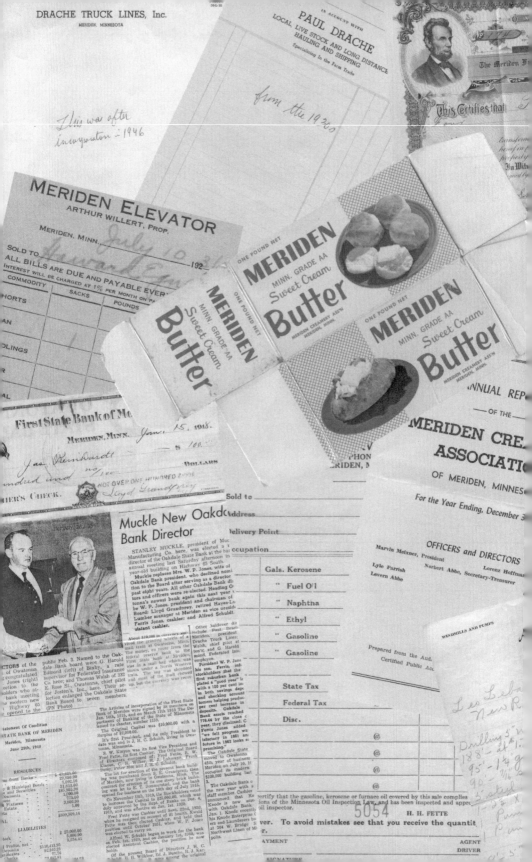

DRACHE TRUCK LINES, Inc.
MERIDEN, MINNESOTA

IN ACCOUNT WITH
PAUL DRACHE
LOCAL, LIVE STOCK AND LONG DISTANCE
HAULING AND SHIPPING
Specializing In the Farm Trade

This was after
incorporation - 1946

from the 192?

The Meriden F...

This Certifies that

MERIDEN ELEVATOR
ARTHUR WILLERT, PROP.

MERIDEN, MINN. July 10 192...

SOLD TO Howard E...
ALL BILLS ARE DUE AND PAYABLE EVER...
INTEREST WILL BE CHARGED AT 1% PER MONTH ON PA...

COMMODITY	SACKS	POUNDS
...ORTS		
...AN		
...DLINGS		
...AL		

ONE POUND NET
MERIDEN
MINN. GRADE AA
Sweet Cream
Butter
MERIDEN CREAMERY ASS'N
MERIDEN, MINN.

ONE POUND NET
MERIDEN
MINN. GRADE AA
Sweet Cream
Butter
MERIDEN CREAMERY ASS'N
MERIDEN, MINN.

ONE POUND NET
MERIDEN
MINN. GRADE AA
Sweet Cream
Butter
MERIDEN CREAMERY ASS'N
MERIDEN, MINN.

First State Bank of Me...
MERIDEN, MINN. Jan. 15, 1918.

$ 100...
Jas. Steinhardt

...ndred and no/100 DOLLARS
NOT OVER ONE HUNDRED D...
...MIER'S CHECK
Lloyd Grandprey CASHIER

Muckle New Oakdale Bank Director

STANLEY MUCKLE, president of Muc...
Manufacturing Co. here, was elected a...
director of the Oakdale State Bank at the ba...
annual meeting last Saturday afternoon in...
year-old building on Highway 65 South...
Muckle replaces Mrs W.P. Jones, wife of...
Oakdale Bank president, who declined nomi...
tion to the Board after serving as a director...
past eight years. All other Oakdale Bank di...
tors and officers were re-elected Heading O...
tonna's newest bank again this next year ...
be W.P. Jones, president and chairman of ...
Board; Lloyd Grandprey, retired Hayes-L...
Lumber manager at Meriden as vice presid...
Farris Jones, cashier; and Alfred Schuldt...
sistant cashier.

OFFICERS and DIRECTORS

Marvin Meixner, President

Lyle Parrish   Norbert Abbe, Secretary-Treasurer
Levern Abbe   Lorenz Hoffman

ANNUAL REP...
— OF THE —
MERIDEN CRE...
ASSOCIATIO...
OF MERIDEN, MINNES...

For the Year Ending, December 3...

Prepared from the Aud...
Certified Public Acc...

...ORS of the
...Owatonna
...congratulated
...Jones (right)
...ction to the
...olders who at-
...bank meeting
...Highway 65
...opened to the

About $18,000 in currency was...
...ered the printing wheels of a m...
...mail train at Owatonna, Minn...
The money, en route from the
...federal reserve bank to the
...First State Bank of Meriden...
...was in a small bag which was
...blown under a North Western...
...train. The bag was shredded
...and most of the mail chewed
...up, but the currency was reco...

...PN Photo)

WINDMILLS AND PUMPS

Les Leba
New R...

Drilling 5"...
188 - 2t 1...
95' - 14 g...
90' - 3....
3 steel...
1-22 X...
Pump...
P...

Sold to ____
...dd Address ____
...er Delivery Point ____
...ccupation ____

	Gals. Kerosene	
	" Fuel Oil	
	" Naphtha	
	" Ethyl	
	" Gasoline	
	" Gasoline	
	State Tax	
	Federal Tax	
	Disc.	

@
@
@
@

5054

...rtify that the gasoline, kerosene or furnace oil covered by this sale complies
...ons of the Minnesota Oil Inspection Law, and has been inspected and appre...
oil inspector.

H. H. FETTE

...ver. To avoid mistakes see that you receive the quanti...
r.

...PAYMENT   AGENT
        DRIVER
        SIGNATURE

DIRECTORS of the
...STATE BANK OF MERIDEN
Meriden, Minnesota
June 29th, 1910

RESOURCES
...from Banks $ 40,021.40
...& Municipal Bonds   200,336.00
...and Securities   7,092.58
...counts   31,013.00
...   180,565.36
...Fixtures   1,500.00
Estate   3,800.00
...   1.00
$509,309.14

LIABILITIES
...ock $ 25,000.00
...   5,000.00
...   3,534.41
...Profits, net $116,415.95
...Deposits   92,360.21
...   301,058.12
$509,309.14

The Articles of incorporation of the First State
Bank of Meriden were signed by 34 members on
Jan. 14th, 1915, and a March 11th 1915 The De-
partment of Banking of the State of Minnesota
...issued its charter, number 116.
The Original Capital was $10,000.00 with a
Surplus of $2,000.00.
It's first President, and its only President to
date was and is J. H. C. Scholdt, living in Owatonna, Minnesota.
S. W. Elmson was its first Vice President and
Fred Fette, first Cashier. The Original Board
of Directors, consisted of Fred Fette, E. W.
Klayne, ... Wilber, H. J. Lohmann, Frank
...berg, Peter Mos and J H. C. Schuldt.
The lot for erection of the present bank building was purchased from E. E. Grandprey, then
...of Meriden, now living in Owatonna, Minn. The
...contract for erection of the one story brick building was let to E. T. Commerford, and the bank
...opened for business on the 16th day of July 1916.
...date was J. H. C. Schuldt.
...On November 11th, 1919, the Stockholders voted
...to increase the Capital to $20,000.00, which was
...duly approved by the Dept. of Banks on Dec. 4,
1919, and was effective on Jan. 1st, 1920.
...Fred Fette was Cashier until Sept. 22nd, 1920,
...when he resigned on account of ill health. Harold
...Fette was then elected Cashier, and held that
...position until October 1921, when W. P. Jones
...was elected to carry on.
...Alfred W. Schuldt began to work for the Bank
...on Feb. 1st, 1926, and on January 1st, 1928, was
...elected Assistant Cashier, the position he now
holds.
...Of the present Board of Directors J. H. C.
...Schuldt, H. H. Willclox, E. J. Borchert, H. J. Kar...

Other holdover di...
include Paul Drach...
Meriden, presiden...
Drache Truck Lines...
ten's; and G. Harold...
Walsh, chief pilot a...
mond, Federated Insu...
employee.
President W. P. Jon...
his son, Ferris, ind...
stockholders that th...
first suburban bank...
ated a "good year" ...
with a 100 per cent in...
in both savings dep...
and checking accoun...
tomers helping produc...
per cent increase in...
deposits. Oakdale...
Bank assets reached...
776.44 by the close ...
year, they disclosed. ...
Ferris Jones added...
"we fell progress was...
isfactory in 1961 an...
future in 1962 looks e...
promising."
The Oakdale Bank...
moved to Owatonna...
Meriden on July 10, ...
occupied its modern...
$100,000 building last...

The Oakdale Bank ...
the new year with a ...
staff member Cashier...
announced that Ken...
Knode is now asso...
Jan. 1. Knode recentl...
his Knode Enterprise ...
ers and Launderers In...
at 304 W. Bridge S...
Northwest Linen of M...
polis.